Antarctica: the next decade

Studies in Polar Research
This series of publications reflects the growth of research activity in and about the polar regions, and provides a means of disseminating the results. Coverage is international and interdisciplinary: the books will be relatively short (about 200 pages), but fully illustrated. Most will be surveys of the present state of knowledge in a given subject rather than research reports, conference proceedings or collected papers. The scope of the series is wide and will include studies in all the biological, physical and social sciences.

ANTARCTICA

THE NEXT DECADE

Report of a Study Group
Chairman
SIR ANTHONY PARSONS
The David Davies Memorial Institute of
International Studies

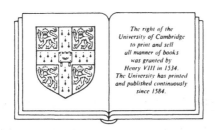

The right of the
University of Cambridge
to print and sell
all manner of books
was granted by
Henry VIII in 1534.
The University has printed
and published continuously
since 1584.

CAMBRIDGE UNIVERSITY PRESS

Cambridge

New York   New Rochelle   Melbourne   Sydney

CAMBRIDGE UNIVERSITY PRESS
Cambridge, New York, Melbourne, Madrid, Cape Town, Singapore, São Paulo, Delhi

Cambridge University Press
The Edinburgh Building, Cambridge CB2 8RU, UK

Published in the United States of America by Cambridge University Press, New York

www.cambridge.org
Information on this title: www.cambridge.org/9780521104036

First published 1987
This digitally printed version 2009

A catalogue record for this publication is available from the British Library

ISBN 978-0-521-33181-4 hardback
ISBN 978-0-521-10403-6 paperback

The David Davies Memorial Institute of International Studies is an unofficial body which
promotes the study of International Relations in all its aspects. It is precluded by its
Constitution from advocating any particular view, or engaging in any form of political
propaganda. The opinions expressed in this book are the responsibility of the authors.

Contents

Acronyms

ASAT   Anti-Satellite
ASOC   Antarctic and Southern Ocean Coalition
ASW   Anti-Submarine Warfare
ATCM   Antarctic Treaty Consultative Meeting
ATCP   Antarctic Treaty Consultative Party
ATS   Antarctic Treaty System
BIOMASS   Biological Investigations of Marine Antarctic Systems and Stocks
BIOTAS   Biological Investigations of Terrestrial Antarctic Systems
CCAMLR   Convention on the Conservation of Antarctic Marine Living
   Resources
CCAS   Convention for the Conservation of Antarctic Seals
DSDP   Deep Sea Drilling Programme
ECOSOC   United Nations Economic and Social Council
EEZ   Exclusive Economic Zone
EIA   Environmental Impact Assessment
FAO   Food and Agriculture Organisation
FCO   Foreign and Commonwealth Office
FOBS   Fractional Orbital Bombardment System
GA   United Nations General Assembly
GLORIA   Geological Long Range Inclined ASDIC
G77   Group of 77
IAEA   International Atomic Energy Authority
ICNAF   International Commission for North Atlantic Fisheries
ICSU   International Council for Scientific Unions
IGY   International Geophysical Year
IIED   International Institute for Environment and Development
IMF   International Monetary Fund
ISTP   International Solar-Terrestrial Physics Programme
IUCN   International Union for the Conservation of Nature and Natural
   Resources

IWC   International Whaling Convention; International Whaling Commission
MAP   Middle Atmosphere Programme
NAM   Non-Aligned Movement
NEAFC   North East Atlantic Fisheries Convention/Commission
NGO   Non-Governmental Organisation
NIEO   New International Economic Order
OAU   Organisation of African Unity
OECD   Organisation for Economic Co-operation and Development
SALT   Strategic Arms Limitation Treaty
SCAR   Scientific Committee on Antarctic Research
SCOR   Scientific Committee on Oceanic Research
SLBM   Sea-Launched Ballistic Missile
SPA   Specially Protected Area
SSBN   Submarine, Ballistic, Nuclear
SSN   Submarine, Nuclear
SSSI   Site of Special Scientific Interest
UNCLOS   United Nations Convention on the Law of the Sea
UNCTAD   United Nations Conference on Trade and Development
UNEP   United Nations Educational Programme
UNESCO-IOC   United Nations Educational, Scientific and Cultural
   Organisation-Intergovernmental Oceanographic Commission
VLF   Very Low Frequency
WCRP   World Climate Research Programme
WMO   World Meteorological Organisation

The Study Group

Sir Anthony Parsons Chairman
Col. Jonathan Alford late Deputy Director, International Institute of Strategic Studies, London
Mr Alan Archer formerly Assistant Director, British Geological Survey
Dr John Beddington IIED/IUCN Marine Resources Assessment Group, Centre for Environmental Technology, Imperial College of Science & Technology, London
The Earl of Cranbrook Member of the House of Lords Select Committee on Science & Technology, the Natural Environment Research Council, and of the Royal Commission on Environmental Pollution
Dr John Heap Polar Regions Section, South America Department, Foreign & Commonwealth Office, London
Dr Martin Holdgate Chief Scientist, Department of the Environment, London
Mr Geoffrey Larminie formerly External Affairs Co-ordinator, Health, Safety & Environmental Affairs Services, British Petroleum, London
Dr Dick Laws formerly Director, British Antarctic Survey
HE Mr Chris Pinto formerly Chairman of the Sri Lanka delegation to the 3rd United Nations Conference on the Law of the Sea
Mr Richard Sandbrook Vice President for Policy, International Institute for Environment and Development, London
Mr Arthur Watts Deputy Legal Adviser, Foreign & Commonwealth Office, London
Dr David Millar Rapporteur

The David Davies Memorial Institute

Miss Sheila Harden Director
Miss Mary Unwin Assistant
Miss Esme Allen Secretary

Preface

In 1985 the David Davies Memorial Institute of International Studies decided to embark on a study of the future of Antarctica. I was invited to chair a Group to produce a book covering the principal aspects of this subject. What follows is the result of our work which has extended over several meetings and a period of about eighteen months.

There is no doubt about the timeliness of this exercise. Many factors are converging in such a way as to oblige governments and other interested bodies to focus their attention on developments in and regarding Antarctica. First, the Antarctic Treaty contains a provision which could open the way for proposals from within the membership for modification or amendment of the Treaty after the expiration of 30 years from its entry into force. This date, 1991, is getting close. Second, an outside challenge to the continuation of the Treaty in its present form has been gathering momentum at the United Nations since 1982. A group of non-aligned States, led by Malaysia, have been canvassing the argument that the present arrangements are too exclusive, particularly in the light of the possibility, however remote, of commercial exploitation of Antarctic mineral and other resources. They are advocating that a more universal system should be negotiated in replacement or extension of the Antarctic Treaty: this would apply to the continent the notion of the 'common heritage of mankind' as enshrined in the Convention on the Law of the Sea, thus bringing Antarctica into line with the concepts of the New International Economic Order.

This initiative has led to a division between the Treaty membership, who advocate the abiding virtues of the present system as it has evolved over the past nearly three decades, and a majority of the Non-Aligned Movement. This disagreement was illustrated for the first time by three split votes on resolutions in the 40th session of the UN General Assembly in December 1985. Thirdly, the non-governmental environmentalist movement has begun to agitate for the reservation of Antarctica for, to use a shorthand expression, 'Greenpeace' purposes.

These cross-currents, which reflect the substantial changes which have taken place within the international community and amongst sections of public opinion since the late 1950s, must be addressed urgently and, if possible, reconciled if Antarctica is to remain immune, as it has been since 1959, from the disputes and dangers which have infected other continents, such as competing national claims, controversies over the exploitation of natural resources and, worst of all, militarisation.

The Study Group worked by allocating chapters to individual members of the Group, circulating the draft chapters, and revising them in the light of the Group's comments. Part I sets out clearly and unequivocally the divergent political viewpoints, while Part II focuses on the functional questions of Science, Living Resources and Conservation, Mineral Resources, and Military Potential. In Part III, I was given the task of attempting, insofar as is possible, to reconcile the differing viewpoints represented in the Group, and of formulating recommendations which could be of value to governments and other interested bodies which may have to face some hard decisions in the years ahead. There were a number of points on which all members could agree, but I take full responsibility for other, more contentious, proposals.

I would like to express my gratitude to all members of the Group, and in particular those who have contributed individual chapters, for devoting so much time to this Study and for making my task as Chairman so easy and agreeable. I would also like to thank Mrs Sally Morphet of the Foreign and Commonwealth Office for her valuable help. And, of course, our collective gratitude goes to the staff of the David Davies Memorial Institute and to Dr David Millar, without whose admirable administrative arrangements and editorial skill the task could not have been undertaken.

Sir Anthony Parsons

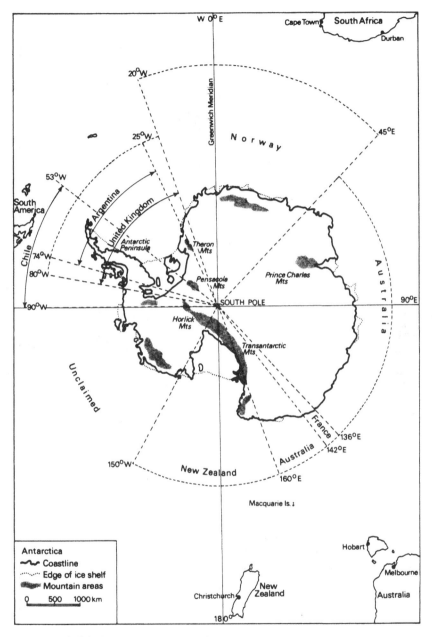

Political map of Antarctica (from Triggs: *The Antarctic Treaty Regime*, Cambridge University Press, 1981)

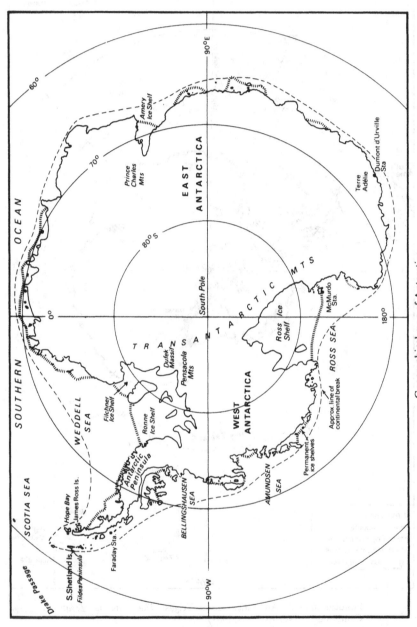

Geographical map of Antarctica.

Part I

The Antarctic Treaty System under stress?

1

The Antarctic Treaty I:
its original and continuing value

In the world of international political affairs Antarctica has been the forgotten continent. Notorious for its harsh climate, renowned for the heroism which characterised its discovery and exploration, appreciated for its spectacular beauty, Antarctica was for most purposes simply there, a standing challenge to human endeavour. Although occasionally the scene of events bringing it to temporary notice, Antarctica's dominant international characteristic was its obscurity.

That obscurity is disappearing. Antarctica is now thought about and discussed not only by States with a direct and active interest in Antarctica but also by States which in the past have never given it more than a passing thought. Each year since 1983 Antarctica has been on the agenda of the United Nations General Assembly, current interest beginning with the statement made in the General Assembly in 1982 by Dr Mahathir, Prime Minister of Malaysia.[1] As a factor in international relations it can no longer be ignored, and as a possible source of friction between States, its future calls for the most careful and responsible consideration.

Antarctica's future is rooted in its present, in the form of the Antarctic Treaty of 1959[2] and the developments which have flowed from it since its entry into force in 1961, and which together now comprise the Antarctic Treaty System (ATS). But that Treaty did not emerge in a vacuum and the immediate background to it is relevant to both its present significance and its future value.

Historical background to the Treaty

In the late 1940s and the 1950s there was considerable international tension in the South Atlantic and the neighbouring parts of Antarctica – tension which led on one occasion to shots being fired, although fortunately only on a small scale.[3] These tensions were mainly the result of disputed claims to territorial sovereignty[4] – Argentina claimed sovereignty over the British territories of the Falkland Islands and part of the Falkland Islands Dependencies (which at that time included what is now the separate British Antarctic Territory, as well as the islands of South Georgia and the South Sandwich Islands), while Argentina and Chile claimed overlapping sectors of Antarctica, those overlapping claims also overlapping in part with the British sector. Four other States – Australia, France, New Zealand and Norway – also asserted sovereignty over parts of Antarctica, although none of their claims overlapped with those of any other. One part of Antarctica, the sector between 90° W and 150° W, was not subject to any sovereignty claims. While these territorial States recognised each other's assertions of sovereignty (except, of course, those States whose claims overlapped), other States which felt a need to take a position in the matter did not recognise them, either on their merits as particular claims to territorial sovereignty or, in some cases, on the broader basis that Antarctica was not in principle capable of being subject to national territorial sovereignty. This complex web of differing views on sovereignty was later to be made even more complicated by the positions adopted by the United States and the Soviet Union which, while not recognising any of the existing claims to sovereignty in Antarctica, maintained that they had a basis for such a claim if they chose to make one. Overall, it is difficult to imagine a situation more likely to promote international tension and discord once Antarctica began to be the scene of increasing activity.

The late 1940s and the 1950s was also one of the more frigid periods of the cold war, and the rivalry and mutual suspicion between the United States and the Soviet Union tended to manifest itself in any part of the world where both powers were present. At that time substantial advances were also being made in weapons technology which, associated with the capabilities demonstrated by the successful launching of artificial satellites (the first Sputnik was launched by the Soviet Union on 4 October 1957), caused a certain nervousness – especially in those States nearer to Antarctica – about the dangerous uses to which the deserted spaces of Antarctica might be put. Notwithstanding such adverse factors, the interests of scientific co-operation enabled the very successful Antarctic programme of the International Geophysical Year (IGY)[5] to go ahead in 1957–58, on the basis of a 'gentleman's agreement'[6] which effectively deprived the governmentally-funded scientific activities of the political implications they might otherwise have had. Against a background which seemed potentially dangerous, this successful neutralisation of political tensions in Antarctica was too good to let lapse, and in 1958 negotiations for what was to become the Antarctic Treaty began in Washington, and were concluded with the signature of that Treaty on 1 December 1959.

What the Treaty has achieved

The Antarctic Treaty was thus the result of diplomacy, building on scientific endeavour. For their own, varied reasons the twelve States who were at the time active in Antarctica and who had participated in the IGY's Antarctic programme, agreed for an unlimited period to put aside their political, military and legal preoccupations in order to establish the Antarctic as an area free for scientific research. They dedicated the continent to peaceful purposes only (Article I); they accordingly demilitarised and denuclearised Antarctica (Articles I and V), and agreed not to press (although, equally, not to surrender) their conflicting views about territorial sovereignty in Antarctica (Article IV); and they opened the way for freedom of scientific co-operation and research in Antarctica on a long-term basis (Articles II and III). They went on to provide (in Article IX) for the essentially scientific and peaceful principles and objectives of the Treaty to be pursued at regular meetings (now known as Antarctic Treaty Consultative Meetings, the States attending them being known as Consultative Parties).

The result was to establish the whole continent of Antarctica as a zone of peace, demilitarised, denuclearised, and subject to an effective system of unannounced on-site inspections, and with the existing problems over territorial sovereignty – such a potent source of international tension – effectively neutralised. The freedom of scientific research established in the Antarctic Treaty, and the obligation to exchange and make freely available scientific observations and results, pointed the way for Antarctica to become an area of unparalleled international co-operation. The twelve original signatories saw that such a Treaty would promote the aims and principles embodied in the United Nations Charter, and it is indeed difficult to imagine a major multilateral treaty more totally in line with the Charter and its broad political and pacific aims.

In 1959 it must have been uncertain how the Antarctic Treaty would fare. Hopes will have been high; realistic expectations may have been more cautious. Now, however, a quarter of a century after the Antarctic Treaty entered into force on 23 June 1961, it can be seen to have been one of the international community's significant success stories. In short, it worked; scientific co-operation in Antarctica has prospered, and as a region of peace in a turbulent world Antarctica has stood out. The Antarctic Treaty has been shown to have served the international community well as a successful, practical and dynamic arrangement.

As a framework for effective international co-operation in Antarctica, the Antarctic Treaty started with many advantages. In comparison with many other treaties of comparable scope, the Antarctic Treaty has the virtues of clarity, simplicity and flexibility. It consists of only ten substantive articles, and its basic provisions on demilitarisation, denuclearisation and freedom of scientific investigation are set out in short and virtually absolute terms. Its organisational structure is simple: instead of any formal international institutions or secretariat, it provides merely for periodic meetings of States active in the Antarctic to consider and consult together on all matters of common interest pertaining to Antarctica. Those States are not only likely to have a direct practical interest in such matters, but are also able to contribute their special knowledge and experience; any other sufficiently interested State can accede to the Treaty, and even become a Consultative Party. It is also noteworthy that although the Treaty provided a diplomatic framework which fostered open scientific co-operation in a political situation which would otherwise have tended to inhibit it, the organisation of scientific work in Antarctica has been left to be co-ordinated through a non-governmental organisation, namely the Scientific Committee on Antarctic Research (SCAR).[7]

By pinning down in categoric terms the crucial elements of the framework for future activities in Antarctica, putting aside the problems to which active pursuit of territorial claims might have led, and establishing open and flexible procedures of consultation and decision-making, the Antarctic Treaty provided a secure base for accommodating the changes which were to come.

Response of the Treaty to changing circumstances

In the quarter of a century since the Antarctic Treaty was concluded, changes in Antarctica have been considerable. Before the Antarctic Treaty (and before the special circumstances during the IGY which preceded it) activity in Antarctica was minimal. The South Pole itself, first reached by Amundsen and Scott in the 1911–12 austral summer, was not again visited on land until the IGY, and is now the site of a year-round scientific station (Amundsen-Scott Station, operated by the United States). In 1986 there were 39 over-winter stations, with more summer-only stations. The 'population' of Antarctica consists now of some 4000 summer personnel, and nearly 1000 who over-winter. There are now two permanent airstrips capable of taking large, wheeled C-130 transport aircraft, and ski-equipped aircraft, even up to the same size, can and do land at many places in Antarctica, including the Pole itself. Whereas even in 1960 the science being done in Antarctica may have had some wider military significance, in the mid-1980s, and in a 'Star Wars' world where military science has global terrestrial and atmospheric dimensions, the possible military applications of Antarctic science are much greater. In 1960 the resources of Antarctica were, apart from whaling, largely speculative; by 1985 whaling had been overtaken in economic importance by the fish and krill resources of Antarctic waters, the exploitation of which is now an annual $400 million industry, while the mineral resources (including hydrocarbons), even though still speculative in commercial terms, have become the subject of inter-governmental negotiations to establish a regime for their development when (and if) that ever becomes a reality. Whereas it was probably assumed in 1960 that expeditions and visits to Antarctica would be essentially government sponsored if not actually government operated, private expeditions have become increasingly frequent, as have tourist visits which are now part of the established scene.

The world outside Antarctica is also a very different place from the world of 1959. One of the most significant changes has been the transformation of the international community which has accompanied the change from the imperial era to the post-colonial era. In 1959 there were 81 States members of the United Nations; by 1986 there were 159. Virtually all of these new States have emerged into independence following the dismantling of the great colonial empires. Although most of these new States are economically disadvantaged they have a substantial political voice. Their advent has generated three important themes in modern international life: suspicion of older international agreements agreed upon before they appeared on the international stage; belief in the principles of universality and 'one-State-one-vote' in the management of the world's affairs; and a desire to change the international economic order so as better to reflect the balance of interests in this new international community. Growing pressures on world resources increase the interest of States, particularly those new States which comprise the countries of the Third World, in the possible presence of major new resources, especially if it is suggested that those resources should remain the preserve of a few States only. A further significant element introduced into the international scene since 1959, and one which cuts across some of the others just mentioned, has been the growing force of environmental considerations, pressed with increasing vigour by those who believe that the environment of our world is precious to all mankind, but is being put increasingly at risk by mankind's actions: the 'Green factor' has become a potent force internationally as well as nationally.

The Antarctic Treaty has proved flexible in evolving in response to changing circumstances, and has shown an increasing appeal. Any State which is a member of the United Nations may accede to the Antarctic Treaty, and any other State may do so with the consent of the Consultative Parties. In addition to the 12 signatory States and original parties to the Antarctic Treaty, a further 23 States have become parties to the Treaty. The full list of parties is: Argentina, Australia, Belgium, Chile, France, Japan, New Zealand, Norway, South Africa, the Soviet Union, the United Kingdom and the United States as original parties, and Czechoslovakia, Denmark, the German Democratic Republic, the Netherlands, Poland, Romania, Brazil, Bulgaria, the Federal Republic of Germany, Uruguay, Papua New Guinea, Italy, Peru, Spain, China, India, Hungary, Sweden, Finland, Cuba, Greece, the Republic of Korea and North Korea which have subsequently become parties. These 35 States account for some 80 per cent of the world's population. Participation includes all the States who are currently active in Antarctica, and all those which are geographically proximate to Antarctica. All five Permanent Members of the Security Council are parties (four were original members, while China acceded in 1983 and joined the others as a Consultative Party in 1985). Parties include the leading members of the Eastern and Western Blocs, leading members of the Non-Aligned Movement such as Argentina and India, and traditionally neutral countries such as Sweden and Finland. In economic terms it associates developing and developed countries, and States with capitalist as well as socialist/Communist systems. Nor is participation in the Antarctic Treaty limited to a particular part of the world: of the 35 current Treaty parties, five are from Latin America, three from Australasia, twelve from Western Europe, five from Asia, one from Africa, seven from Eastern Europe, and one each from North America and the Caribbean. In geographical terms, the only major absentees are Arab and African States (the one member from Africa being South Africa). It has thus become a Treaty which associates powerful and less powerful States across the broad range of international interests, and with an extensive geographical spread. The increased participation in the Antarctic Treaty was not simply the result of an early enthusiasm inspired by the Treaty's novelty. Indeed, accessions have increased in recent years, rather than the reverse; Treaty membership doubled between 1981 and 1984.

Similarly, the Treaty did not establish the Consultative Parties as a closed group. It is based, in its origins as well as in its subsequent development, on the involvement of those States which have a sufficiently active interest in the subject matter of the Treaty – a principle which underlies many successful international arrangements, whether they involve regional groupings of a primarily political or economic nature, or common functional concerns such as the management of a regional fishery or the control or prevention of regional pollution problems. Article IX.2 allows for any acceding State to become a Consultative Party if it 'demonstrates its interest in Antarctica by conducting substantial scientific research activity there, such as the establishment of a scientific station or the despatch of a scientific expedition'. Six of the eighteen present Consultative Parties have become such in that way: Poland became a Consultative Party in 1977, the Federal Republic of Germany in 1981, Brazil and India in 1983, and China and Uruguay in 1985.

Establishing a station or dispatching an expedition are only illustrative criteria: the real test is the conduct of 'substantial scientific research activity there'. While establishing a permanent station (as have the six States named in the previous paragraph) may well be an expensive undertaking, it is not a necessary requirement in order for a State to become a Consultative Party, and indeed it is something which it may not even be in the general interest to encourage, since it leads in practice to relative over-crowding of the most easily accessible parts of Antarctica, with a number of attendant disadvantages. Other forms of scientific research activity, which might in any case be just as desirable and valuable, may not be so expensive and need not, necessarily, be beyond the resources of countries which have a serious interest in Antarctica. It is that degree of interest which, rightly, comes first, with participation in Consultative Meetings following as a consequence, reflecting that interest. The evident capacity of the Antarctic Treaty to evolve suggests that the requirements of Article IX need not exclude States with a serious practical interest in Antarctica from becoming Consultative Parties. The fact that this involves some expenditure for prospective Consultative Parties, not necessarily excessive or disproportionate for those seriously interested in participating in Antarctic affairs and the management of the continent, is not in itself unreasonable. It reflects the realities that, simply by reason of its geographical circumstances, Antarctica is not a place in which it is easy to be active, and that few international activities are cheap and none is free. While there is room for judgement in assessing whether a State has conducted substantial scientific research activity in Antarctica, it is significant that the test for a State wishing to become a Consultative Party, and thus have a decision-making role in Consultative Meetings, is a demonstrated capacity to contribute knowledge about the Antarctic. The test is internationally unique, and is conducive to the general good.

The expansion of international participation in the Antarctic Treaty has been matched by a steady increase in the range and content of discussions at Consultative Meetings. At those meetings the Consultative Parties not only consult together on matters of common interest pertaining to Antarctica, but also recommend to their governments measures in furtherance of the Treaty's principles and objectives. Consultative Meetings are in practice normally held every two years, the most recent – the thirteenth – in 1985. By the end of that meeting 154 recommendations had been adopted, covering the whole range of activities in Antarctica.[8] From the start these recommendations have dealt with a wide range of issues, such as help in emergencies, telecommunications, historic monuments, logistics, scientific co-operation and tourism. A growing proportion of recommendations has been devoted to the conservation of Antarctic wildlife and to the prevention of damage to the Antarctic environment. Binding instruments have been adopted for the Conservation of Antarctic Flora and Fauna,[9] the Conservation of Antarctic Seals,[10] and the Conservation of Antarctic Marine Living Resources.[11] These instruments and recommendations and the Treaty itself, together with the various measures – national and international – taken pursuant to them, constitute a comprehensive regime for Antarctica and form the Antarctic Treaty System[12] – a formally separate but in fact closely related network of agreed arrangements and commitments.

In this way, Antarctica today is subject to a legal regime which ensures its effective and peaceful administration in a manner consistent with the purposes and principles of the United Nations Charter. A combination of national and international measures and instruments ensures that there is no juridical vacuum in Antarctica. This clearly distinguishes Antarctica from outer space, or the deep sea bed, where new international arrangements – based on the emerging concept of the common heritage of mankind – were created in order to fill what would otherwise have been a gap in the area of legal regulation. (It may in passing be noted that any application to Antarctica as a whole of 'common heritage' concepts would involve disregarding the territorial sovereignty in Antarctica exercised or claimed by seven States.[13])

The Antarctic Treaty's capacity to evolve in order to meet new circumstances and interests has been one of its strengths, and is well illustrated by the emergence of possible Antarctic mineral resources as a subject of growing importance. During the negotiation of the Antarctic Treaty the suggestion was made that, in order to forestall the serious disputes which could follow from the discovery of valuable deposits and which would prejudice the wider purposes of the Treaty, effective provision should be made for dealing with a future possibility that mineral wealth might be found in Antarctica. It was felt at the time that this possibility was too remote, and could raise too many serious difficulties, to make it desirable to try to cover the matter in the Antarctic Treaty itself. But it was not long before the issue again surfaced, and discussion of it fell within the general powers of Consultative Meetings. Discussion in Consultative Meetings started in the early 1970s. By 1981 the stage had been reached where it was agreed to be appropriate to begin negotiations for a regime to regulate the exploration for, and development of, Antarctic minerals resources.[14]

By that time the Consultative Parties had had practical experience of the value of the Antarctic Treaty, and of the scientific co-operation which it fostered, and had invested considerable political capital in the Treaty's continuation. The potential of the minerals issue to disrupt the Treaty was, however, not diminished. While the States asserting sovereignty could, under Article IV of the Antarctic Treaty, agree not to press their views in the context of the general 'peaceful purposes' provisions of the Treaty, and its encouragement of scientific co-operation, it would be a very different matter for them to agree under that Article that mineral wealth could be removed from Antarctica as if the claims to territorial sovereignty did not exist. The future discovery of mineral wealth could put a strain on Article IV – the cornerstone of the Treaty – which it might well not be able to resist, with consequential damage to the Treaty structure as a whole. Gradual increases in scientific knowledge of the area, and the effects of the 1973 oil crisis together with the renewed speculation about 'new' sources of energy which it prompted, did nothing to encourage Governments confidently to believe that the hypothetical dangers which could be foreseen would in the event never materialise. Even the severe technological and economic difficulties in the way of commercially justifiable extraction of mineral resources in the unfavourable circumstances of Antarctica did not offer satisfactory reassurance, since the possibility of events outside Antarctica making it politically necessary to press ahead in the search for Antarctic minerals on a basis other than just prospective commercial profitability could not be completely discounted.

However, the fact remained that, at the time (and still), no commercially exploitable mineral deposits in the Antarctic were known to exist, nor were there (or are there now) immediate plans to begin exploration for them. The potentially serious risks to the Antarctic Treaty system were accordingly seen to be most appropriately avoided by attempting now to negotiate a minerals regime the aim of which was not to encourage mineral activities in Antarctica or to establish a mining code to regulate them in detail, but rather to provide an agreed framework within which, *should they ever occur*, they could be carried out in a manner which would take account of the various interests involved; high among these interests was the need to avoid future conflicts in the course of what might otherwise be an anarchic scramble for resources, and to ensure full safeguards for the Antarctic environment. These negotiations are still in progress.

That they are taking place at all, at a time when no commercially exploitable mineral resources are known to exist in the Antarctic, and their development (and even exploration for them) is still a very long way off, further illustrates the Antarctic Treaty System's capability and readiness to anticipate and resolve a new issue before pressures to begin development become severe, vested national interests appear, and the risks of conflict become greater. It is an exercise in prudent forethought, which must be preferable to waiting until it has to be a matter of crisis management in circumstances where the traditional way of settling the underlying disputes about territorial sovereignty would be by the use of force.

The evolution that has taken place in the range of subjects discussed at Consultative Meetings has been matched by change in the nature of those meetings. They have increasingly taken on the character of an international governing body for Antarctica, while increased attendance at Consultative Meetings, growing outside interest, and the possibility of a review of the Treaty's operation after another five years, have made those meetings steadily more political. Virtually all aspects of human activity in Antarctica are now discussed at Consultative Meetings, and many are regulated to a greater or lesser degree by measures adopted at them. In recent years there has also been something of a shift in emphasis, in that scientific and associated technical matters have increasingly to share the agenda with institutional and constitutional matters touching on the better management of the system as a whole.

Attendance at Consultative Meetings has also changed. While originally restricted to Consultative Parties, as foreseen by Article IX of the Antarctic Treaty, it was nevertheless found both possible and desirable to allow those acceding States which were not sufficiently actively involved in Antarctica to qualify as Consultative Parties to attend Consultative Meetings as observers, with the right to participate in discussions[15] – and in practice they do so in a manner indistinguishable from that of the Consultative Parties themselves (although decision-making, by consensus, is still restricted to the Consultative Parties). Because the Antarctic Treaty has no normal institutional structure, the role within it of the Consultative Parties and non-Consultative Parties may be seen as in some respects an amalgamation in one forum – Consultative Meetings – of roles that would in other circumstances be played in two separate institutions: discussion of items at Consultative Meetings has something of the character of a General Assembly, in which all parties may participate, while for decision-making the meeting has something of the character of an executive organ with limited membership. While no two international institutions have identical structures, since each adopts a structure suited to its own functions and circumstances, the Antarctic Treaty, in evolving a system incorporating both general and limited membership elements, has followed the broad pattern common to the great majority of international organisations.

The original justification for such a structure remains. This is that those States which have demonstrated a practical involvement in Antarctica are best qualified to take decisions affecting conduct in Antarctica, decisions which are almost all concerned with limitation of their freedom of action there. Such a principle is familiar in many international institutions, where States particularly qualified by experience or involvement have a special role in the appropriate decision-making organ. It is a principle whose application is particularly compelling in relation to Antarctica, where physical conditions are uniquely difficult, the scientific activities which form the bulk of the conduct which is under discussion are sophisticated and complex, and the political framework established by the Antarctic Treaty is very finely balanced. It is also very relevant that, unlike most international organisations, substantive decisions at Consultative Meetings are effectively taken by consensus,[16] which in effect gives any State participating in decision-making a veto. While this may sometimes make it more difficult to reach agreement, and may slow the pace at which development may take place, it has the compensating advantage that those agreements that are reached command the support of all Consultative Parties, and are therefore more likely to be observed. These consensus arrangements are workable in a context in which participants have a direct and active involvement in Antarctica, but could be less so where some participants might have no such direct stake.

Prospects for the future

The framework for Antarctica provided by the Antarctic Treaty was established for an unlimited period (although there is a possibility of review after the Treaty has been in force for 30 years).[17] There seems no reason why it should not continue to provide an appropriate basis for the management of international activity in Antarctica. As experience over the last quarter of a century has shown, the Antarctic Treaty is capable of adapting its procedures and practices in such a way as to cope efficiently with the many new circumstances which have arisen. Although in many ways the Antarctic Treaty System has now grown well beyond the Antarctic Treaty which was its origin in 1959, that Treaty continues to be the firm foundation for the whole system. Its capacity for internal adaptation to new circumstances, always on the basis of consensus among the Consultative Parties, has been one of its greatest strengths in the past and is a critical safeguard for the future of Antarctica. The Antarctic Treaty represents a very delicate balance between otherwise competing interests, and unless Antarctic developments take place within the framework of the fundamental elements incorporated in the Treaty the risks of destabilisation are considerable.

To jeopardise the Treaty's achievements would be an act of folly. It would not, moreover, be merely a matter of putting at risk the achievements of the Antarctic Treaty, but also the whole interlocking series of instruments and measures which constitute the Antarctic Treaty System of which that Treaty is the key component. There would follow serious risks of resurrecting those dangerous elements which were present before the IGY – the possibility of the arms race extending to Antarctica, its possible use for nuclear purposes, the strident revival of territorial claims, and perhaps the addition of new ones. Tempting though the thought may be, it is naïve to think that it would be an easy matter, or even possible at all, now to re-negotiate the Treaty keeping only some elements which meet with wide acceptance and jettisoning the rest. The Antarctic Treaty was negotiated at a time when circumstances were propitious for concluding an agreement reflecting the delicate balance of interests which was required, and the States concerned made the most of that opportunity. Circumstances now are very different, and there can be no confidence that a similar package, reflecting a similarly acceptable balance of political, strategic, scientific, and legal elements, could be concluded today.

Fortunately, although the Antarctic Treaty System can doubtless, like most things, be improved – and is indeed continually improving by adapting to new needs – its demise is probably remote. The broad international reality is that politically and militarily the present arrangement in the Antarctic serves the (different) political interests of all the major participants. In particular, it appears to be satisfactory for the two superpowers, whose influence in maintaining a *status quo* which they support should not be underrated. It is, however, not only the participants in the Antarctic Treaty who find that its continued existence serves their interests. For many other States the Treaty has great value in preserving the whole continent of Antarctica in a sort of political and strategic limbo; it is, in a world otherwise full of tension, one large area which they can for the most part forget about. The southern flanks of the Indian, Pacific and Atlantic Oceans are protected by, in effect, a non-nuclear zone of peace underwritten by the superpowers. It is a state of affairs which many States, even though not parties to the Antarctic Treaty, are – rightly – reluctant to disturb. Even many of the States which in the United Nations debates since 1983 have found reason to look critically at certain aspects of the Antarctic Treaty System have acknowledged its value, and have disavowed any intention of seeking its destruction.

Sight should also not be lost of the important fact that the Antarctic Treaty is a treaty lawfully concluded by sovereign States directly concerned with its subject matter. *Pacta sunt servanda* is an accepted fundamental principle of international law, reflected not only in the Vienna Convention on the Law of Treaties but also in the preamble to the UN Charter itself.[18] While third States normally do not derive international rights and obligations from the treaties concluded between other States, they equally have no legal basis for calling into question the validity of such treaties. The Charter of the United Nations may be regarded as superior law where obligations under another treaty conflict with obligations under the Charter,[19] but in the absence of any such conflict – and there is no such conflict in respect of the Antarctic Treaty – obligations under that other treaty stand, and must be observed and respected. For States, either individually or acting collectively through such bodies as the United Nations, to seek to undermine the lawful treaties of other States is to embark on a course which can put at risk much more than the particular treaty under present attack.

Conclusions

The Antarctic Treaty System, fully consistent with the principles and purposes of the United Nations Charter, and effectively promoting their attainment in the area with which it is concerned, has, since its inception and at least until very recently, been accepted by the international community as part of the general structure of international relations. It has kept a continent free of international disputes, and free for international co-operation. It has shown itself to be in practice an effective and dynamic international legal instrument which has served the international community well. In showing the way in which international co-operation may be fostered in circumstances which at first sight might seem calculated to encourage conflict rather than co-operation, and in revealing the extent to which States, by mutual forbearance from taking their differing views to logical and possibly mutually inconsistent conclusions, can develop a spirit and practice of peaceful co-operation, the Antarctic Treaty stands as one of the major achievements of international co-operation in modern times.

2

The Antarctic Treaty II:
the case for change

From the viewpoint of those States which are not parties to the Antarctic Treaty, the international community in a wider sense, the situation regarding Antarctica often appears to be much less satisfactory than the protagonists of the Antarctic Treaty System portray it to be. For some non-member States, the realisation that, after more than five thousand years of recorded history, some one tenth of the planet is not subject to any generally recognised comprehensive legal regime is cause for disquiet, particularly when the rate at which technology continues to enhance man's destructive and exploitative capabilities is borne in mind. Such a lacuna in the fabric of recognised governmental control and management, occurring as it does in an area remote from the usual centres of human activity, and one moreover of critical importance to the maintenance of ecological balance over the planet as a whole, would seem to offer scope for mischief of a grave order.

While federalists may find the division of the world into nation-States to be productive of selfish, even destructive, trends, such a division does, at least, enable the greater part of the world to be governed, and to be subjected to broadly similar systems of law and law enforcement which, taken together, offer a reasonably secure context for the conduct of peaceful human activity. The icy wastes of Antarctica alone remain isolated from the mainstream of international legal, social and economic development. As we approach the twenty-first century a large portion of the Antarctic remains the object of unresolved claims and counterclaims by a handful of States, and the non-recognition or outright rejection of those claims by others. The remaining portion of the continent appears to lie in a kind of legal limbo – unique as the last land territory on the planet to which no State has yet laid claim.

Importance of the Antarctic and its resources

Much has been written in recent years of the natural resources of the Antarctic and its surrounding seas. There are those who see the Antarctic as the repository of enormous reserves of minerals, ranging from uranium to oil, and others who dispute that assessment, or who argue that the technology required for the safe and economic recovery of these resources is not likely to be developed in the foreseeable future. There are those who see the waters around the Antarctic as a vast storehouse of food capable of meeting all of the world's protein needs for years to come, and others, more cautious, who emphasise the dangers of ill-considered and excessive exploitation. It is, however, generally agreed that the Antarctic and the surrounding waters are of critical importance in influencing the planet's climate – a fact of far-reaching significance for all living things.

One central fact, expressly acknowledged in the Antarctic Treaty of 1959, is surely beyond debate: the Antarctic and its resources are of immense importance, actual or potential, to all mankind, and to every State in the community. It is against this background that the future of the Antarctic ought to be considered, and contemporary efforts at co-operative action in the area evaluated.

The Antarctic Treaty of 1959

Between July 1957 and December 1958 twelve countries operated research bases in the Antarctic in connection with the International Geophysical Year. The reluctance of some participants to dismantle their bases and to leave at the end of the IGY seems to have led to the negotiation of a *modus vivendi* which sought to reconcile, presumably on a temporary basis, the interests of two groups of countries: the claimant States on the one hand, and the non-claimant researching States on the other. The result was the Antarctic Treaty.

The Treaty itself is of narrow scope and of possibly limited duration, since it allows for review after 30 years of operation. It prohibits measures of a military nature, including the testing of any type of weapon, or nuclear explosions, or the disposal there of radioactive waste material. It provides for 'freedom of scientific investigation . . . and co-operation toward that end . . .' not of unlimited scope, but rather investigation and co-operation of a kind 'applied during the International Geophysical Year'. Claims to parts of the Antarctic are suspended, and activities during the pendancy of the Treaty would have no role in the assertion, support or denial of any claim, nor may any new claim be asserted while the Treaty is in force. This Treaty, it must be observed, in common with other such treaties, can bind only the States parties to it – 12 at its creation, and 32 in mid-1986.

States not parties to the Treaty have viewed with interest the apparent intention of the parties to enlarge the scope of the Treaty, and to transform their commitments to co-operation for scientific research into a complex of management obligations with respect to living and non-living resources, establishing these obligations through a number of inter-linked and supervised ancillary treaties, combined with certain recommendations to the membership. This wider view of the Treaty is sometimes referred to as the Antarctic Treaty System. States not parties to the Treaty have paid little heed to these developments, bearing in mind the general rule that a treaty creates neither obligations nor rights for a third State without the latter's consent. However, recent initiatives taken by the Treaty parties – barely one-fifth of the membership of the United Nations – have economic implications which may engage the economic and wider interests of the community as a whole, interests which are expressly recognised and preserved by the Treaty parties.

Evolving role of the Consultative Parties

At the heart of the Antarctic Treaty System are the Consultative Parties. These are the original signatories named in the preamble to the Antarctic Treaty, plus other States which have demonstrated their interest in Antarctica through 'conducting substantial scientific research' there, and currently numbering some eighteen States in total. According to Article IX of the Treaty, member States confer on the Consultative Parties a 'steering' function, which includes the making of specified types of recommendations to governments of States parties of 'measures in furtherance of the principles and objectives of the Treaty . . .' (Articles IX.1 and IX.4). The making of such recommendations is subject, however, to *unanimous* approval by the Consultative Parties – a valuable safeguard of essential political interests through the power of veto that is rare in a modern political context. Other powers conferred on the Consultative Parties include the right by unanimous agreement to modify or amend the Treaty (and therefore, presumably, to extend or terminate it as well) (Article XII), to inspect any part of the Treaty area through observers designated by them and to make aerial observations of such areas (Article VII), and to invite non-members of the United Nations to accede to the Treaty (Article XIII.1). Any Consultative Party may initiate a conference to review the Treaty, and the support of a majority of Consultative Parties is required for the adoption of any proposed revision of the Treaty (Article XII.2). The unanimous recommendations of the Consultative

Parties would appear to be binding on themselves. They are expressed in imperative or mandatory terms and, by an evolving practice, the Treaty parties seem to recognise in the Consultative Parties a quasi-legislative role since they automatically accord 'approved recommendations' a measure of effectiveness in relation to current and new Treaty parties, and even *erga omnes*.[1] The Final Report of the Fourth Consultative Meeting also declares that:

> 2. Recommendations which become effective in accordance with Article IX of the Treaty are, in terms of that Article, 'measures in furtherance, of the principles and objectives of the Treaty '.

This would seem to amount to recognition of the Consultative Parties' right to interpret the Treaty, albeit for Treaty parties only; any subject on which the Consultative Parties adopt measures unanimously is, by definition, to be considered to be within the scope ('principles and objectives') of the Treaty. Even States not parties to the Antarctic Treaty may be required to recognise in the Consultative Parties a 'responsibility for the protection of the Antarctic environment from all forms of harmful human interference', should they become parties to the Convention on the Conservation of Antarctic Marine Living Resources.[2]

The 'principles and objectives' of the Treaty also seem subject to an evolutionary process. Article IX, paragraph 1(f) of the Antarctic Treaty empowers the Consultative Parties to recommend to States parties measures in furtherance of the 'preservation and conservation of living resources in Antarctica'. Recommendations under this provision have led to the establishment of new international agreements developed essentially among the parties, but opened for signature to States which are not parties (Convention for the Conservation of Antarctic Seals (1972), Convention on the Conservation of Antarctic Marine Living Resources (1980)).

While the Treaty foresees a role for the Consultative Parties in making recommendations regarding the conservation of living resources, this is not the case in relation to non-living resources. It is significant, therefore, that measures to regulate Antarctic mineral exploration have been formulated through a series of Special Meetings of the Consultative Parties, begun in 1972, the eighth of which convened in 1986. This initiative would appear to complete the entry of the Treaty parties into areas of regulatory activity related to the management of Antarctic resources, and thus to underscore the evolution of principles and objectives of definite economic significance. This capacity of the Antarctic

Treaty System to evolve, frequently mentioned with warm appreciation by the Treaty parties during debates in the United Nations, may well be at the root of sharpening differences between them and the overwhelming majority of States outside the Treaty framework, for the most part developing countries, seeking to bring about a New International Economic Order.

Of the developing countries, Argentina and Chile, as claimant States, are among the original signatories of the Antarctic Treaty, and six others – Brazil, Cuba, India, Papua New Guinea, Peru and Uruguay – have, for different political reasons, subsequently become parties to the Treaty. These States must be taken to support the Treaty, and to be able to reconcile its structure and functions with the New International Economic Order to which they are committed. On the other hand, attitudes to the Treaty among the great majority of the developing countries seem to range from the indifferent and sceptical to the outspokenly critical. Recently, on the initiative of a group of States led by Malaysia, attention has been focused on fundamental questions regarding the Antarctic and its resources, and upon the need for co-operative measures to resolve them based on principles of common interest and universal participation.

The Antarctic Treaty and the New International Economic Order
Inter-dependence
An essential element of the New International Economic Order (NIEO) is the concept of inter-dependence. Inter-dependence, as a relationship between States, is seen both as a goal to be achieved, and the reality that should motivate States toward that goal. Inter-dependence, and two of its practical expressions – the sovereign equality of States, and the universal participation of States in economic decision-making – are of importance in assessing views of the Antarctic Treaty held by States not parties to it, and are thus of relevance when considering the future of arrangements concerning the continent and its resources.

To the developing countries, with relatively weak economies and decades behind the industrialised countries in their capacity to acquire and apply the technologies needed to speed up economic development, the recognition that States form a 'world community', and of 'the reality of inter-dependence' referred to in paragraph 3 of the NIEO Declaration[3] is logical enough. So is the recognition, referred to in the same paragraph, that 'international co-operation for development is the shared goal and common duty of all countries'. To those who have a history of 'depen-dence', 'inter-dependence' is a concept easily grasped and, in a way, reassuring. But neither as 'goal' nor as 'reality' is inter-dependence attractive to the industrialised countries, and in particular, to those which have assumed far-flung military obligations in response to perceived threats to the security of their economies, political systems, and ways of life. To them, economic *independence* is both the goal and the reality, and 'inter-dependence' is a concept too closely associated with a hungry, inept, yet stridently importuning crowd, demanding no less than the politically impossible – the adjustment of developed country economies in a manner that could impair the quality of life of their populations – and resulting in a withdrawal of popular support from the governments in power. For them the concept of inter-dependence is better retained in the realm of philosophy, except, of course, in time of war or other national emergency – such as the prospect that some commodity essential for industrial growth (such as oil) might become subject to control by a cartel of the normally dependent. In that event, it might be prudent to emphasise the 'reality' of inter-dependence for as long as might be necessary.

Sovereign equality of States and universal participation

A fundamental aspect of inter-dependence, and one which may be described as legislative or managerial, is expressed in two sub-paragraphs of paragraph 4 of the Declaration establishing the NIEO:

(a) Sovereign equality of States . . .

. . .

(c) Full and effective participation on the basis of equality of all countries in the solving of world economic problems in the common interest of all countries . . .

The Charter of Economic Rights and Duties of States adopted by the General Assembly later in the same year contains a similar provision:[4]

Article 10

All States are juridically equal, and as equal members of the international community, have the right to participate fully and effectively in the international decision-making process in the solution of world economic, financial and monetary problems

. . .

These texts give expression to a basic principle currently supported by the overwhelming majority of States, and one which seemed of no less than universal validity before the breaking of colonial ties in the 1960s trebled the original membership of the United Nations: the principle whereby each State, regardless of size or other attribute, is recognised as having only one vote of a value equal with that of every other State, decisions being made by a simple or qualified majority of participating States. Inevitably, this principle seems to cause its adherents to lean away from acceptance of forms of decision-making which would give preponderant effect to the vote of a particular State (or group of States) on the grounds that it had some special interest in or expertise concerning an issue, or had made a greater financial or other contribution, or was first in the field. While systems of weighted voting were incorporated in the Bretton Woods agreements, and the power of veto was claimed by the permanent members of the Security Council in the political climate prevailing immediately after the Second World War, such devices have generally been rejected when proposed during more recent negotiations of general international agreements.

The Antarctic Treaty of 1959, and all the instruments created within the Treaty system, conferring as they do special powers and responsibilities on the Consultative Parties, appear to that extent to give effect to a principle that is today widely regarded as unacceptable. The status of a Consultative Party, which under the Treaty would exercise evolving powers with clear implications for the exploitation of natural resources, and thus with the potential to bring about economic consequences for States generally, may only be achieved either through an historical association with the Treaty (by having participated in the Antarctic programme of the IGY), or through a demonstrated interest in Antarctica by conducting substantial scientific activity there. The first requirement, by definition, excludes all but a handful of States, all of which are already parties; the other is outside the technological and financial capacities of most States.

Trends in the negotiation of general multilateral treaties over the last quarter of a century make it unlikely that the granting of special decision-making powers based on financial and technological capacity – as provided for in the Antarctic Treaty – would be acceptable to the overwhelming majority of States today. Negotiations regarding other areas of economic and strategic importance, such as outer space, the moon and other celestial bodies, and the sea-bed beyond national jurisdiction, have led to treaties providing for decision-making on a 'one-State-one-vote' basis. The Outer Space Treaty creates no 'steering committee', nor is a State required to qualify for membership by, say, placing a satellite in orbit or a man on the moon. Similarly, the demonstrated capacity to carry out deep sea-bed mining is not required for membership of the Council of the International Sea-bed Authority, nor does it confer special voting rights on a member. On the contrary, while a variety of economic considerations (and not merely financial capacity) are relevant in the election of some members of the Council, half the membership is elected on the basis of geography alone.

In the case of the Convention on the Law of the Sea, during the negotiation of provisions for voting in the Council of the International Sea-bed Authority, and also those of the Governing Board of the Enterprise, efforts were made on several occasions to replace the one-State-one-vote/majority-vote decision-making mechanism favoured by most States, with some systems which would give decisive weight, or a power of veto, to States expected to make larger financial contributions toward institutional and operational costs. None of these systems proved acceptable, and the Conference was able to achieve a near-consensus on the basis of one-State-one-vote.

Application of the one-State-one-vote principle does not, of course, imply that there should be no 'steering' body of limited membership, but rather that such a body should be constituted and subject to regular renewal in a manner that would ensure the equitable representation of interests, and that the principle of one-State-one-vote would apply to decisions made by that body. An acceptable one-State-one-vote system which combined rules for the protection of minority rights might be devised, for example one that required that issues be decided by qualified majorities (the majority required in each case increasing in magnitude in accordance with the level of 'importance' ascribed to the issue), and which was applied by a body constituted and operated so as to ensure equitable representation of different interests among the membership.

Under the Antarctic Treaty, the requirement of unanimity (Article IX.4) and its corollary, the veto for each Consultative Party, is an extreme form of protection of minority interests which would be hard to incorporate in arrangements that emphasise community interests. By its nature it is more suited to arrangements which seek to reconcile individual States' interests.

The Antarctic Treaty as reconciling individual States' interests

Several types of interests are reflected among the current membership of the Antarctic Treaty, an individual member being likely to have more than one concern. Thus, a State may have an interest in scientific research, and look upon Antarctica as a unique laboratory. The search for knowledge may, however, also be an expression of an interest in discovering new and independent sources of minerals required for industrial use, or of an interest in developing new technologies associated with mineral extraction. A State may have an interest in a study of the environment, which may or may not be related to its mining and processing interests. Other States with global military interests may find a detailed knowledge of the terrain and conditions of the Antarctic essential to their overall strategies. There are, in addition, some seven claimant States with interests in seeking recognition of their claims to sovereignty over parts of the Antarctic and their resources and, in the case of claimant States located relatively close to the Antarctic, with national security interests as well.

The 1959 Treaty brought these individual State interests into balance. Since then, inaccessibility of the continent, a lack of interest in it on the part of non-member States, and the careful exercise of discretion and diplomacy by the Treaty parties has contributed to the maintenance of that balance for over two decades. Claimant States, in return for not having their sovereignty claims actively disputed by the few States which might have found overriding strategic reasons for doing so (and which also possessed the military and technological capacity to ignore those claims with impunity), were content to suspend their claims as far as those States were concerned. On the other hand, for States with strategic, and possibly industrial, interests in the area and its resources, it was convenient to have sovereignty claims suspended so as to have the freedom of the continent, at the same time underwriting the balance achieved, and making common cause with the claimants in relation to non-members, whose occasional interference could invite awkward questions about their own activities, no less than the legitimacy of territorial claims. The Treaty 'system', given direction by its Consultative Parties, and with its evolving fabric of recommendations, practices and understandings, seemed the surest guarantee that the balance would be maintained, perhaps even more than the legal content of the Treaty itself did. New members, and in particular, new Consultative Parties, joining with the knowledge of the Treaty's internal balance, and finding it useful, would be unlikely to challenge it.

What has now begun to focus attention increasingly on the Antarctic Treaty arrangements, and which may even lead to the eventual replacement of the existing balance of interests, is the expression of a new category of interest in the area and its resources. Whereas the Antarctic Treaty brought about a balance of individual State interests (albeit sometimes represented as having been inspired by the 'interest of all mankind'), the group of States led by Malaysia at the United Nations has now articulated a 'community interest' in the Antarctic, and effectively demonstrated it by securing some 92 votes in favour of a resolution calling for the international management and equitable sharing of the benefits of

Antarctic minerals, as well as enhanced involvement of international organisations in the study of the area and its resources.

Community interest and policy 'differential'

Negotiations connected with regimes designed to give effect to a 'community interest' have in the past proved to be very different from those connected with resolving bilateral issues. When a 'community interest' is the basis of a national policy urged at a conference, considerations of equity may result in agreement on waiver, in certain circumstances, of a strict reciprocity as between industrialised and developing countries. The UN Convention on the Law of the Sea, for example, contains several provisions on preferential treatment for developing countries in such areas as preservation of the marine environment, marine scientific research, and the transfer of marine technology.

Negotiations on the 'common heritage' concept at the UN Conference on the Law of the Sea may serve to illustrate the policy elements associated with the presentation of individual (or like-minded group) interests on the one hand, and of community interests on the other. The developing countries as a group tended to emphasise their perception of 'community interest' as fundamental to their position, while among the free market economy industrialised countries, individual economic and security concerns appeared to dominate. Table 2.1 attempts to compare some policy elements characteristic of the two approaches (but not those of the socialist industrialised countries).

The policies of the industrialised countries derive from internal economic and political situations requiring that their governments respond effectively and without delay. While they may take into account their perception of the 'community interest' in the resource, the most pressing duty for a representative of an industrialised country at a conference would be to ensure that the expressed requirements of a constituent, such as an industry on which the health of the country's economy depends, should be met. The policies of the developing countries, on the other hand, are often not the result of internal pressures from powerful constituents, but depend rather on the views of the leadership as to what is best for the country. Such imperatives are likely to be aimed at ensuring management of the resource so as to postpone its exploitation until the States concerned are capable of participating in it on an equal footing; or, if the resource were to be exploited immediately without their direct participation, at ensuring that they would suffer no adverse economic effects, and that their rights in it would be realised through the receipt of benefits in due proportion to the interest they perceive in the

Table 2.1

Industrialised	Developing ('community interest')
1. Need for raw materials for industry, to maintain economic and social standards as well as to pursue military objectives	1. Need for raw materials for reconstruction and development of weak economies to raise economic standards to minimum level; apprehensions regarding decline in land-based mineral export earnings
2. Technology for competitive exploitation of resources available with private industry	2. No technology for competitive exploitation, and difficulties of acquisition
3. Basic belief in free market/open regular choice of form of government	3. Basic belief in dirigistic approach/open or controlled choice of government
4. Industrial corporations as constituents influential in choice of government, press for maximising competitive advantage in resource exploitation	4. No immediate demand from constituents requiring articulation of particular sea-bed mining policies
5. Government policies intended to maximise competitive advantage for constituents, and to protect parent free market system:	5. Government policies intended to establish benefit-sharing and to prevent decline in export earnings from land-based mining, or postpone exploitation, to offset lack of capacity to compete:
(a) freedom of access to resource	(a) controlled access to resource
(b) minimal regulation of operations	(b) strict regulation of operations
(c) first come, first served (*prior in tempore, potior in jure*)	(c) selection of operator by reference to community benefit promised
(d) exploitation activity not prohibited deemed permitted	(d) activity not expressly permitted must be negotiated
(e) freedom of scientific research, within minimum conditions imposed on commercially-oriented research	(e) scientific research encouraged, subject to control aimed at: (i) preventing its use as cover for subversive or military activity, (ii) ensuring wide dissemination of information concerning data, (iii) equitable distribution of benefits from commercial use of data
(f) contribution from each according to capacity, with benefits and management power accruing in proportion to contribution (risk)	(f) contribution from each according to capacity, distribution of benefits according to need, with equitable representation of main interests on management body, and one-State-one-vote
(g) institutional decision-making: emphasis on protection of minority rights through a rule of consensus.	(g) institutional decision-making: emphasis on majority rule, with safeguards for minority positions through consensus-oriented procedures.

resource. It might be said that their position is aimed at protecting their future or contingent interests.

The fact that policies of the industrialised countries may be formulated in response to immediate and even urgent internal economic and political concerns, while the developing countries seem to be taking up positions aimed at preserving their rights in the future, thus bringing the groups into a kind of inter-temporal conflict, should not be taken into account when assessing the validity of either of them. Both positions are taken with a view to dealing with real and important problems. A government's policy aimed at resolving the problems of constituents does not necessarily have a superior claim to validity merely because those problems are of an immediate nature, particularly where implementation of those policies has the potential to infringe the rights of others. A policy is not necessarily inferior or expendable merely because it is intended to protect contingent interests, or to provide compensation if an interest is to be eroded or taken away altogether. Policies oriented toward the conservation of species of animals or plants, or protection of the environment, are of no less validity or practical importance merely because they are designed to preserve the rights of future generations in certain assets available today.

It would not seem fair, therefore, to suggest, as some Treaty parties have done, that the community interest articulated by Malaysia and States not parties to the Antarctic Treaty, is an interest in what should be to them a 'distant' or 'theoretical' issue, implying that it should be left to be dealt with by those who are 'closer' or who have a more 'practical' interest, i.e. the Treaty parties, and in particular the Consultative Parties. Malaysia's initiative focused attention on the Antarctic, just as Malta's initiative in 1967 compelled all the members of the United Nations to examine the full implications of permitting unregulated exploitation of the resources of the deep sea-bed. Given the community's undoubted interest in both subjects, it would seem reasonable to acknowledge that any State has not merely the right but also the duty to initiate public debate and decision with respect to them. While the regime for sea-bed mining that emerged might not be acceptable to a few, it can hardly be denied that the implications of exploiting sea-bed resources were widely discussed, and that a system of regulation which appeared to meet the interest of the community was agreed among the overwhelming majority of States acting in an informed and co-operative manner.

At a treaty-making conference a wide variety of policies of more or less equivalent 'validity' and 'importance' need to be given due consideration. It can no longer be the simple objective of a delegate to use the means at his disposal – be it the economic and military influence of his State, or the votes of the majority he is able to rally to his cause – to advance his government's policies at the expense of those of his opponents; it is his task rather to undertake an informed co-operative effort to resolve the problems created by inconsistent policy positions, so that the essential needs of all are met as fully as possible. This, it would seem, is the modern approach to the international legislative process – an approach which recognises equivalence in the weight of each legislator's opinion, and derives the force of an emerging rule from general acceptance of a reasoned reconciliation of those opinions.

The refusal by one group of States to take due account of the concerns of another group, discounting 'the reality of inter-dependence', could lead to Pyrrhic victories. The developing countries, through insisting on a regime which reflects their perception of 'community interest', but which is rejected by those who own or control the technology needed to exploit the resource, would go on to produce complex legal structures which are of little practical value. On the side of the industrialised countries, technological independence alone may not, it seems, be sufficient to secure them the benefits of the resource when faced with the unrelenting disapproval of the majority of States. Both attitudes merely result in confrontation, and postponement of the operations which are the avowed objective of each regime.

These would appear to be lessons to be learned from the UN Conference on the Law of the Sea. The industrialised countries stand in varying degrees of reluctance to bring into force the sea-bed mining provisions adopted by the Conference, and broadly acceptable to the developing countries. The developing countries, on the other hand, have through the so-called 'moratorium resolution'[5] and successive statements of a similar kind,[6] served notice that they are unitedly and firmly of the opinion that mining of the deep sea-bed which is not carried out under the terms of the Convention – even if undertaken unilaterally under domestic legislation, or in the context of a treaty between 'reciprocating States' – would be in breach of international law. While lawyers may question the 'legally binding' character of these statements, their value as evidence of the existence of an *opinio juris* has not been lost on the investor on whose co-operation the financial viability of sea-bed mining – and indeed of any such high-technology, high-risk operation – must ultimately depend, and to whose hard-earned dollar a host of more secure investments beckon.

Restructuring the Antarctic Treaty System

The resolution adopted by the General Assembly at its 40th session as Resolution 40/156 amply demonstrates the existence of a community interest in Antarctica, and indicates in addition that some two-thirds of the organisation's member States believe that current arrangements under the Antarctic Treaty are not an adequate expression of that interest. While Part B of the resolution affirms that 'any exploitation of the resources of Antarctica should ensure . . . non-appropriation and conservation of its resources and the international management and equitable sharing of the benefits of such exploitation' (operative paragraph 1), it stops short of pre-judging the conceptual framework within which action might be taken to achieve this end. Malaysia seemed to contemplate a new phase of Antarctic activity in which the United Nations would have a role in a system of management characterised by 'accountability, involvement, and equity' – which may be interpreted to mean accountability of the managers, involvement of all those who wish to participate (without preconditions of any kind), and the sharing of benefits on the basis of equity. Such a system would seem entirely consistent with the essentials of decision-making on world economic issues to which the developing countries have hitherto given their unqualified support. A restructuring of Antarctic Treaty arrangements may be implied. However, as it stands, the proposal does not seem to be confined within any particular ideological limits, and should be explored 'without prejudice' by all concerned.

It may not be useful or desirable to disturb the balance of individual States' interests currently achieved under the Antarctic Treaty. The reciprocal commitments based on political expediency, which hold together parties as radically opposed as the claimant States and those who totally reject their claims, are powerful indeed. At their centre are the Consultative Parties, each of whom currently enjoys the protection of what is today a rare device: the power of veto. As with splitting the atom, the problem may well be that of controlling any reaction that would follow interference with the forces which maintain the current balance. It might therefore be best to undertake the delicate task of building upon the balance achieved under the 1959 Treaty by the addition of other elements until a new balance is generally agreed upon. Work toward creation of a new and generally acceptable balance should take place under the auspices of the United Nations, and may take into account the following possible approaches:

1. Essential undertakings provided for under the 1959 Treaty may be maintained, including the prohibition of measures of a military nature, and of nuclear explosions and the disposal of radioactive waste, the maintenance of freedom of scientific research, the suspension/non-recognition of territorial claims for the duration of the Treaty, and procedures

for becoming a party to the Treaty. Others concerning resource management may be added so as frankly to acknowledge the Treaty's new scope and avoid the need for strained interpretations of the present text or recourse to 'implied powers'.

2. The concept and role of Consultative Parties under the 1959 Treaty may be retained. However, the composition of the group may be altered to provide for more representative participation by the entire membership in decision-making: (a) those States which are currently Consultative Parties should be assured of their continuation in that role in recognition of their special interests in the area; (b) any State which in future conducts 'substantial scientific research' in the Antarctic should be admitted as a Consultative Party as at present provided; (c) a new provision should be added, expanding the number of Consultative Parties so that they comprise: (i) the current Consultative Parties, (ii) any new Consultative Party admitted on the basis of 'substantial research', and (iii) as many other members elected for three-year terms by the Treaty parties as may be necessary to ensure that the total number of Consultative Parties is equally divided between Consultative Parties under (i) and (ii), and others,[7] in such a manner as to ensure equitable geographical representation among the Consultative Parties as a whole;[8] (d) if the number of Treaty parties does not permit full implementation of such amendments, a transitional article could provide for the closest approximation; (e) each Consultative Party would have one vote, and devices for safeguarding minority interests, such as qualified majorities for certain decisions, should be added. The unanimity rule should be replaced by a 'consensus' rule, and its application restricted to decisions of an exceptionally sensitive nature negotiated and defined in the Treaty.

3. Consideration may be given to other constitutional changes such as the following: (a) provision for subsidiary organs of the Consultative Parties, which would regularly review (i) legal and organisational questions, and (ii) scientific and technical questions, and make recommendations thereon. The latter would take over the functions currently performed by SCAR, and might be constituted so as to give effect only to those changes in SCAR which appear necessary to avoid inconsistency with the general trend represented by the changes effected; (b) provision for institutionalised relationships with the United Nations (Economic and Social Council or, if this is not acceptable, the Security Council), and its subsidiary organs and specialised agencies (UNEP, WMO, FAO, UNESCO-IOC) and regional commissions, as well as the International Sea-bed Authority when it comes into being, the better to give effect to the provisions of Article III, paragraph 2 of the Antarctic Treaty, which

requires that 'every encouragement shall be given to the establishment of co-operative working relations with those Specialized Agencies of the United Nations and other international organizations having a scientific or technical interest in Antarctica'. The Consultative Parties should undertake responsibility for the regular dissemination of information on the Antarctic throughout the UN system; (c) provision for periodic review of Antarctic arrangements, say every ten years, at a conference attended by the Treaty parties as full participants, and by all other members of the United Nations as observers.

4. Such suggestions frequently give rise to the criticism that they imply the creation of an unwieldy and costly bureaucracy. Acceptance of the principle of universal participation is, of course, likely to imply the establishment of an organisational structure more complex than that required in order to administer exchanges among a restricted and cohesive group of States. On the other hand, permitting such a structure to become unwieldy or disproportionately expensive is a matter which the membership responsible for its financial support may deal with effectively in the light of its priorities. The practice of establishing the level of a member State's contribution to the budget in proportion to its financial capacity has in recent times given rise to serious problems. Major contributors, upon failure of the organisation to implement their policy priorities, have withdrawn or threatened withdrawal, leaving extensive programmes without financial support, or forcing restrictions on their scope. It may well be time to consider a different approach, and to require instead a uniform level of contribution to an institution's regular budget, fixed so as to be within reach of any State. No participant would be permitted to contribute more than its share. The positive aspects of such an approach are that it would be consistent with a one-State-one-vote procedure for decisions, and it would restrict the size of any secretariat to the minimum. On the other hand, it would severely curtail or eliminate altogether an organisation's capacity to become involved in operational programmes. While the latter might be financed on a voluntary basis by the more affluent States, the terms and conditions on which such operations would be carried out would need to be the subject of decisions by the membership as a whole.

5. If changes along the lines of those referred to in the preceding paragraphs are achieved, the Consultative Parties may be given an expanded and more detailed mandate covering, for example, the regulation of resource exploration and exploitation on the continent as well as its offshore areas.

6. Consideration may be given to conferring a special legal status on the unclaimed sector of Antarctica and its offshore areas, perhaps making it, as a transitional measure, the responsibility of the United Nations, and including a representative of the United Nations elected by the General Assembly at meetings of Consultative Parties (*cf*. Namibia).

Conclusions

The time seems ripe for a fundamental reappraisal of arrangements for Antarctica, with the Antarctic Treaty parties playing a leading role. Unfortunately the Treaty parties appear united in interpreting the initiative taken by the developing countries at the United Nations as an unwarranted and ill-conceived 'attack' on a system which, they insist, is of proven efficacy. In response, they have chosen merely to reiterate the virtues of current arrangements and refuse to participate in votes taken on the item. Massive non-participation, such as was seen during the votes on resolution 40/156, may be understandable in the light of initial resentment at developing country persistence, but is hardly a means of achieving a reasonable reconciliation of views expressed in the debate, or of preventing Antarctica from becoming the scene or object of international discord.

One may expect that, after a period of reflexion, other counsel among the Treaty parties might prevail, and that they may find it possible to join the great majority of States in looking with fresh eyes at the Treaty of 1959 in order to determine what adjustments need to be made to it 'in the interest of all mankind'.

3

The Antarctic Treaty III: non-governmental organisations, conservation and the environment

For many non-governmental organisations (NGOs) the current interest in the future of the Antarctic has been a wonderful opportunity. Where else can one at the same time question concepts of sovereignty, security, international equity, the environment and the rule of international law, without regard to indigenous people? Where else can the untouched beauty of the Earth be presented so evocatively, and the opportunity for political adventurism, or even mischief making – all in the name of ideals – be so easily secured? Yet when all the initial rhetoric of the campaigners is peeled away what remains is not entirely clear, or straightforward. It is easy enough for them to call for 'internationalising the area', 'creating a World Park', and 'opening up the Antarctic system', but when the implications of such demands are understood in terms of continued demilitarisation, non-exploitation or regulated exploitation of resources, and modifications to the sovereignty claims, then their demands become mixed up with the realities of political power and, as a result, become less clear.

Contrary to popular belief many of the demands of the NGOs are hardly radical. Greenpeace International, for example, does not call for the abolition of the Antarctic Treaty,[1] and the International Union for the Conservation of Nature and Natural Resources (IUCN) has modified its original demands for a World Park in Antarctica. When faced with the representatives of governments the leaders of most organisations appear to be open to argument and compromise. There is no doubt, however, that NGOs, of many colours, will continue to confront governments with hard choices for a protected environment, and this confrontation is going to continue for as long as it seems to be good for the environment. The pattern of demand followed by concession followed by demand has continued for ten years now, and shows no signs of abating.

Objectives of the NGOs

What exactly are the demands that the NGOs have been pressing? Broadly speaking their objectives, which are often common with those of the non-Treaty nations, fall into five categories:

1. *Public accountability* – there should be adequate information, easily obtainable, on Antarctic activities and meetings.
2. *Non-discrimination* – the Antarctic regimes are not universal. They work against States that cannot mount major scientific programmes in Antarctica, thus creating two tiers of membership.
3. *International equity* – in the event that benefits arise from Antarctica, they should be widely shared within the international community, and involve the participation of a wider community in the decision-making process.
4. *Conservation* – Antarctica's (as yet inadequately understood) role in world climate, its value to science as a pristine area, and its aesthetic values, are such that they should be adequately protected in perpetuity.
5. *Living resource conservation* – the Convention on the Conservation of Antarctic Marine Living Resources (CCAMLR), with its consensus decision-making, should be modified so that the aspirations of Article II of the Convention are realised.

Looking at the concerns of specific organisations, Greenpeace has stated:[2]

> While recognising that it has some weaknesses, *Greenpeace supports the Antarctic Treaty system* [emphasis added]. There are several aspects of the Treaty which are extremely good which could not be renegotiated in the present climate of international relations. Greenpeace believes that one of the major weaknesses of the Treaty System is that it has yet to gain acceptance as the appropriate forum for Antarctic decision-making from the majority of the world's nations. One of the obstructions to gaining truly international recognition is the apparent exclusiveness of the Antarctic Treaty System and, in particular, the secrecy which prevents other interested States (and non-governmental organisations) from having any formal input. States which are unable to afford major scientific programmes in the Antarctic are excluded from any meaningful participation in decision-making. It is Greenpeace's hope that the discussion of Antarctic issues in such fora as the United Nations will lead to a better understanding of the needs and aspirations of States

outside the system by Consultative Parties. The Treaty System should be prepared to adapt itself in response to these types of discussions to remedy weak points if and when they become apparent.

Later, at the time of the Special Consultative Meeting on Antarctic mineral resources in September 1985, Greenpeace set out their position more precisely:[3]

> Greenpeace believes that the Antarctic, including its mineral resources, should be protected. The following fundamental principles should be applied to all human activity in the Antarctic:
> * There should be complete protection for all wildlife.
> * Protection of the Antarctic wilderness should be paramount.
> * The Antarctic should remain a zone of limited scientific activity, with co-operation between scientists of all nations.
> * The Antarctic should be a zone of peace, free of nuclear and other weapons and all military activities.
> These principles collectively encapsulate the idea of the Antarctic as a 'World Park'.

Greenpeace went on to suggest to Antarctic Treaty members that they permanently prohibit mineral activities in the Antarctic by establishing a 'protection regime' which would give effect to the World Park concept, maintain the existing moratorium on mineral activities pending agreement on a protection regime, and establish an Antarctic Environmental Protection Agency with a mandate to regulate any activity taking place within the Antarctic Treaty area that might have a detrimental environmental impact – including science, logistic support and tourism.

Whilst it is difficult to demonstrate empirically that Greenpeace is the most outspoken advocate of change to the *status quo*, it can hardly be considered as being in the back seat. Yet even Greenpeace is apparently not arguing for an outright revolution in Antarctica.

Gradual modifications of the Treaty processes have been at the heart of the NGO positions of late, to be achieved by work with the parties to the Consultative Meetings or by way of demonstration and public pressure without. At the end of the day their posture appears to be one of negotiated or reasoned change, and not outright abolition of the Treaty System that the Treaty members fear. However, it is also very clear that any refusal by the Treaty members to hear and to reason with their critics could result in an upsurge of strong rhetoric or mischief on the part of some organisations. In effect, the Consultative Parties have it in their power to prevent or to provoke a stronger reaction from the heretics.

They will have to face up to the pressure that the NGOs, often in informal alliance with the governments of non-member States, bring. It is for this reason that the modifications demanded by such organisations are very much a part of the mix that will determine the future of the area. The Treaty members have to make a decision; do they want a useful dialogue with the NGOs, or do they want confrontation on the issues they raise? At present, as the Treaty parties continue with the minerals regime negotiations, it appears that some of the NGOs have grown weary with consultation; others have never begun it. For example, in April 1985 the Consultative Parties decided not to invite representatives of any international organisations, whether intergovernmental or non-governmental, as observers to the 13th Consultative Meeting. Similarly, the Antarctic and Southern Ocean Coalition (ASOC) and various other individual NGOs have never succeeded in gaining access to meetings arising under CCAMLR. Thus Greenpeace (Auckland) stated in 1985 (in an internal newsletter):[4]

> Throughout our involvement with CCAMLR, we have been the very models of decorum, and have done nothing in the way of direct actions. If we continue to be politely fobbed off by the majority of member nations, and CCAMLR is unable to enforce the conservation measures so desperately needed in the Southern Ocean, then perhaps a different approach is called for.

Recent concessions to accommodate non-Treaty views

In order to evaluate the nature of the pressure for change within the Antarctic Treaty System, let us consider firstly the concessions which have been achieved to date, by pressure from a mixture of NGOs and non-Treaty States, and secondly those that might be achieved in the future by the same means. They can be divided into 'institutional' and 'operational' issues.

Institutional issues

Since 1982–83, when the current interest in Antarctica first arose in the United Nations and other international forums outside the Antarctic Treaty System itself, the following steps toward 'opening up' the ATS have been taken.

In the matter of increased participation and institutional development, the Consultative Parties have sent invitations to acceding States to attend Consultative Meetings as observers (12th and subsequent meetings, 1983 onwards), and to attend the minerals regime negotiations (1984 onwards).

At the 12th Consultative Meeting a mandate was given to invite representatives of international organisations with relevant scientific or technical expertise to attend future Consultative Meetings as observers (however, this mandate was not acted upon at the preparatory meeting for the 13th Consultative Meeting in April 1985). Invitations were also extended to SCAR and to the CCAMLR Commission to attend the 13th Consultative Meeting in 1985, and to appoint observers to present reports on developments within their respective areas of competence at the 14th Consultative Meeting in 1987. This latter point is important because it forced the Consultative Parties to amend the rules of procedure to allow participation by observers from international organisations, thus opening the door to participation by additional such observers in the future. However, there has in practice been some reluctance to invite such observers, and the subject is due to come up again in preparation for the 14th Consultative Meeting. Since 1983, Australia, New Zealand and Denmark have followed the American practice of including representatives from non-governmental organisations on their delegations. SCAR is also considering establishing a new 'associate' membership category, and may invite non-member countries of SCAR to designate observer participants at its 19th meeting. In addition, other States have joined the Antarctic Treaty System; since June 1983, China, India, Hungary, Sweden, Finland and Cuba have acceded to the Treaty, and Brazil, India, China and Uruguay have become Consultative Parties. Further accessions and applications for Consultative Party status are expected. Amongst those who are currently active or who are planning future expeditions are Italy, Sweden, Spain, the Netherlands, Finland and Peru.

Regarding the NGOs' demands for information to be more readily available, new procedural steps ensure that representatives of acceding States are provided with the relevant documentation prior to Consultative Meetings and the minerals regime negotiations. At the 13th Consultative Meeting the American government also indicated that, as the depositary government for the Antarctic Treaty, it would assist all contracting parties to the Treaty in obtaining documentation from previous Consultative Meetings at which they were not present. Delays to date in acting on these decisions seem to stem more from insufficient attention to this problem than deliberate exclusion, but nevertheless the acceding States were in some cases deliberately excluded from document circulation during the minerals regime negotiations. At the 13th Consultative Meeting steps were taken to amplify the report of the meeting, and to ensure that it provides continuity of information from one Consultative

Meeting to the next, including information on follow-up actions taken by States and other relevant bodies.

Finally, various steps were taken in the period 1983–85 to widen the circulation of the reports of Consultative Meetings, the Antarctic Treaty Handbook, and the annual exchanges of information. Certain documents were declassified from previous meetings (documents from the first three Consultative Meetings are no longer confidential), and information on where materials on Antarctica may be obtained by interested parties was also made available. SCAR is reviewing its publications in order to see how they respond to information needs and requests on Antarctica.

Operational issues

Since the inception of the Antarctic Treaty forum, Antarctica has been viewed as a 'special conservation area' by those responsible for its management, but the increased public attention given to Antarctica in the 1980s has stimulated change. The following additional steps have been taken to address environmental protection and conservation measures (not including steps taken by CCAMLR).

The application of environmental impact assessment (EIA) procedures to Antarctic science and logistics activities has been considered. While several countries already require such assessments by law, others have no established practice in this area. At the 13th Consultative Meeting, parties were unable to agree on binding procedures, but a number of them indicated that they would abide by the procedures included in the report[5] (this issue is to be considered again at the 14th Consultative Meeting, as well as the value of increased comparability between EIA procedures). At the 12th Consultative Meeting, the Code of Conduct governing waste disposal for Antarctic expedition and station activities was given inconclusive reconsideration with a view to revision, and will be discussed again at the 14th meeting. There is now a requirement for improved reporting on tourist activities, and on the impact they may have on Antarctica's pristine values (13th meeting). There has also been a call for consultations between Antarctic programme managers, with the objective of avoiding problems due to overcrowding or concentration of station activities.

In 1985 three new Specially Protected Areas (SPAs) and thirteen new Sites of Special Scientific Interest (SSSIs) were established and eight SSSIs were extended. However, no agreement has yet been reached on the establishment of marine SSSIs, a long-sought objective of the NGO community. There has been a call for the study of an additional category of protected area, and of steps to improve the comparability and accessi-

bility of scientific data on Antarctica, with a view to long-term management and conservation needs.

These last two steps are potentially significant since they would provide
the basis for long-range planning and management of Antarctica and
activities there. The data collection and management issue is a critical first
step in developing this plan. Such material would also greatly enhance the
quality of a regional conservation strategy, an idea that the IUCN (in
co-operation with members of SCAR) is working on.

Future objectives that the NGOs may press
Institutional issues

In general, the environmental NGOs want the long-term planning for Antarctica to be deliberately designed so as to consider environmental protection and conservation aspects. Many also want the implications for international participation in the Antarctic Treaty System more
fully explored. In effect, can an accommodation of the wider community
be found without losing the values of the ATS, and can the distinction
between Consultative Parties and non-Consultative Parties be reduced?

To allow both to happen, deliberate steps would have to be taken by the
Treaty parties in order to plan for an expanded programme of international participation in Antarctic science by those who are interested.
SCAR could play a role, as could other international organisations, and
the merits of shared programmes and logistics facilities should be
explored. It is also suggested by the NGOs that the Consultative Parties
expand the opportunities for interested foreign nationals, particularly
those from non-Consultative Party nations, to participate in their national
Antarctic research programmes on an individual basis.

Greenpeace remain active in pushing for a World Park in Antarctica
and plan that four of their people should overwinter in the 1987 season.
The team will consist of a doctor studying the effects of the cold and
confinement, a marine biologist monitoring wildlife particularly under
the pack ice, and two others to run the camp and help with wildlife
observation. In addition there are plans for the Greenpeace boat to
monitor other Antarctic bases during the year. But science is not the
primary objective. It is political. By maintaining a non-governmental
presence they hope to highlight the alternatives open to Treaty governments. It is, in short, classical pressure group politics and although
extremely ambitious, not to say risky, is no different in concept from the
protester chained to the proverbial railings.

The NGOs will doubtless continue to press for observers from relevant
international organisations (including non-governmental ones) to take

part in Consultative Meetings, in order for the Consultative Parties to develop more extensive working relationships with them. It is argued that these relationships would mutually benefit the ATS and the observer organisations through the exchange of information and expertise. They would result in a greater measure of (indirect) participation in Antarctic policy formulation by universal membership organisations and, through expanding the dissemination of information on the ATS, would build confidence about the Treaty System amongst outsiders. The international organisations in particular could be helpful in the planning and implementation of broad-based co-operative international science programmes, thereby saving the expense of each individual nation establishing its own facilities.

This approach would also result in the Consultative Parties having to consider the implications of shared facilities and programmes for consultative status under the Treaty. Needless to say, if there were a move toward more co-operative programmes, then the level of co-ordination and harmonisation of data between the different Antarctic regimes would become more important. With this and other points in mind, NGOs have argued for the establishment of a Treaty secretariat. This could contribute a number of important functions, all of which would tend to open up the system. These functions might include the maintenance and distribution of information in response to outside interest, the co-ordination of activities between the different Antarctic forums and institutions, reviewing the annual information exchanges (with a view to increased co-ordination between science programmes), defining opportunities for expanding international co-operation in Antarctic science, identifying potentially damaging activities and the need for environmental impact assessments of them (including the receiving and circulation of EIAs and comments received), and identifying potentially beneficial monitoring programmes that could be easily combined with proposed annual activities at little extra cost (not least as a means of filling gaps in the data important to the long-term conservation of Antarctica).

Finally, the Consultative Parties may have to develop a formal mechanism to enable the concerned public to contribute their views on Antarctic matters. Obviously, designation of observers and NGO participation on delegations helps, as would the circulation of critical documents and proposals for comment prior to final decisions.

Criticisms of ATS information practices have grown more muted during the past year, and this trend is likely to continue as the States involved take further steps to ensure the availability of existing information. Nevertheless, in addition to continuing the trend of amplifying

reports of Consultative Meetings, circulating more widely (and declassi-
fying) the meetings documents and the annual exchanges of information,
the following specific steps would continue the more open policy: the
making available to all interested parties of the draft text of the mineral
regime negotiations, and pursuing consultations on the negotiations with
such outside parties; making available the inspection reports produced
under the ATS (and maintaining them as part of the public record);
ensuring that SCAR documents are automatically made available to
Treaty parties that are not members of SCAR; and making available
documentation from the Special Consultative Meetings (after they are
concluded) to interested parties as part of the public record. A national
reporting function on activities taken to implement and enforce ATS
legislative and practical measures should be established. This would
supplement reports of Consultative Meetings (to the extent that they
address these questions); similar reports should also be called for under
the minerals regime.

Conclusions

The non-governmental organisations will not leave Antarctica
alone. The minerals regime negotiations ensure their continued interest,
and a diverse range of countries from the Netherlands to Malaysia will
give them an audience. The concessions already made by the Treaty
members have gone a long way to overcome past difficulties. Indeed, had
more account been taken of NGO suggestions at the time of the
CCAMLR negotiations for majority voting and better information
reporting many of that agreement's current difficulties would have been
avoided. From the Treaty point of view one is tempted to suggest that
further concessions are the way to stem the tide of discontent that could
follow.

But perhaps the mere negotiation of a minerals regime is too great a
provocation to the non-Treaty parties and NGOs alike? One suspects so;
once the spectre of a share-out is raised, the level of interest in the Treaty
is significantly increased. As Barbara Mitchell has commented:[6]

> The Antarctic Treaty does not seem to provide a basis broad and
> solid enough upon which to construct resource management
> regimes. But it remains a unique and valuable instrument which
> has prompted scientific research, protected Antarctica's environ-
> ment from the advent of man and his weapons and ensured
> peaceful co-operation in the area, keeping territorial conflict at
> bay. It should be preserved and more countries encouraged to

join. The antagonism currently expressed towards the Treaty would be dispelled if it were made clear that it did not claim to provide the constraints for controlling the area's resources. If anyone is trying to undermine the Treaty it is the Treaty parties themselves by asking too much of it.

There is no reason why Treaty Parties should not enjoy an important position in a future mineral regime, but it would be wiser for everyone to approach these negotiations from a different direction, recognising the interests of the wider community and trying to clarify the legal basis for operations. If the objection is raised that this would violate Article IV of the Treaty, it would be worth considering to what extent the draft minerals regime in effect violates this accommodation by divvying up the different attributes of sovereignty.

The Treaty is a vital, if limited purpose instrument. The best guarantee of its survival is to recognise these limitations.

Part II

Uses of Antarctica

4

Science

The Antarctic amounts to nearly one-tenth of the land surface of our planet and a tenth of the world ocean. This significant geographical extent, if for no other reason, calls for research activities, but the general case is strengthened by the following considerations. It is the highest and coldest continent by virtue of its polar position, which results in all but 2 per cent of the land surface being covered by ice (over 4 km thick in places). It was the 'keystone' of the former supercontinent of Gondwana, and hence of great interest in terms of plate tectonics. As the strongest cooling centre of the global system it has importance for meteorological and climatic studies. Its isolation from the other continents by a wide and deep ocean is in part responsible for the fact that it is still relatively unaffected by man, and therefore provides a baseline for studies on global pollution of various kinds. Finally, a number of biological problems can be better studied there than anywhere else, and a vast, unplanned, long-term perturbation experiment caused by the over-exploitation of the southern stocks of whales provides further unique opportunities to study interactions within a marine ecosystem.

An additional advantage of the Antarctic for scientific research is the unparalleled political situation which exists there: the absence of national boundaries under the Antarctic Treaty helps to promote large-scale international co-operation in scientific programmes. However, the high costs of supporting research in such a remote and rigorous environment still means that there need to be very good reasons for conducting the research there rather than elsewhere, although the sharing of logistic support tends to encourage inter-disciplinary programmes. There is a wide diversity of specialised fields of science which provide opportunities for exciting and novel work of academic, strategic and applied value. These more specific research problems are described below.

Geology and geophysics

Serious research in the earth sciences in Antarctica has only been undertaken for a little more than a quarter of a century, so our geological knowledge of the region is limited. The fact that only 2 per cent of the land area is exposed rock, the hostility of the environment, and the difficulties of logistics are significant additional handicaps. The earth sciences are of increasing academic importance, and there are major strategic and applied problems to be tackled.[1,2] The Antarctic Treaty parties are currently formulating proposals for a minerals regime (which will also need to take account of sciences other than geology and geophysics if the potential environmental impact of commercial activities is to be given due attention).

Greater (or East) Antarctica is for the most part a structurally complex Precambrian shield, which is important as an accessible (in the sense that it lies at the earth's surface, albeit ice-covered) and very ancient part of the earth's crust, over 3500 million years old. The major rock exposure in Antarctica is the Transantarctic Mountains, and this feature forms a natural geographical boundary between Greater Antarctica and the younger Lesser (or West) Antarctica. In contrast to Greater Antarctica, Lesser Antarctica consists of a number of micro-continental blocks separated by subglacial basins.

The research fields involved include structural geology and plate tectonics, sedimentology, palaeontology and stratigraphy, marine geology, radiometric dating, Precambrian and Phanerozoic history, radio echo-sounding, magnetometry, palaeomagnetism, seismic sounding and gravity surveys. Vast subglacial basins separate the sparse outcrops where conventional geological studies are possible, and these studies are therefore complemented by airborne remote sensing, although this is largely limited to radio echo-sounding of ice thickness and airborne magnetometry. Synthetic aperture radar could help to distinguish differing bedrock types, and satellite remote sensing is also important. At sea, complementary marine geophysical and marine geological research is carried out on the continental shelves and ocean basins, when and where the pack ice distribution permits.

There are several problems in plate tectonics of global significance that are being investigated in the Antarctic: of these, the break-up of the original supercontinent of Gondwana is of major interest. The present Antarctic plate is bounded by seven other major lithospheric plates. In the reconstructions that have been made, the Pacific margin from the Antarctic Peninsula to New Zealand is the region where understanding is most imperfect. The importance of such studies is that they may point to the reasons for the break-up, and so help us understand major processes within the earth's mantle.

An important and related problem concerns the nature of the junction between Greater and Lesser Antarctica. This involves questions about the extent, structure, geological history and relationship of the microcontinental blocks within Lesser Antarctica, and their relation to the single mass of Greater Antarctica. In addition to this fundamental problem there is great interest in the nature of the intervening basins and their potential as hydrocarbon reservoirs; the solution of these problems will require seismic traverses. South America, Africa and Australia all possess rocks containing metallic minerals which are of actual or potential value as resources, and in Antarctica the Dufek Massif and the Transantarctic Mountains contain similar minerals.

Another outstanding problem in reconstructions of Gondwana is that most of them require the overlap of the Antarctic Peninsula with South America. This is related to the opening of the Weddell Sea, which is thought to have occurred slowly over a time scale of some 170 million years, and also to the growth of the Scotia Sea to the north. The latter provides excellent examples of ridge crest–trench collisions as the sea developed by sea-floor spreading over a period of 30–40 million years. Studies using remote sonar sensing (such as GLORIA (Geological Long-Range Inclined ASDIC), Sea Mark or Sea Beam) would help to identify spreading centres and fore-arc fabrics, and possibly the collisions, to assess the dynamics. They would be coupled with seismic investigations, dredging, coring and heat-flow studies, and are expected to yield exciting results.

The western margin of the Antarctic Peninsula is a region of great importance where subduction-related processes are being studied. Such studies have practical implications; the metallic minerals exploited in the Andes are related to subduction along their western margin, and it seems likely that there is a similar relationship in the region of the Antarctic Peninsula. This is unusual among ancient arc systems in that there is good exposure of the terrains, and having been active over 180 million years it is also one of the longest-lived magmatic arcs in the world. The record of subduction over the last 65 million years is extremely well defined. There are other opportunities for studying marginal basins and fore- and back-arc basins in this region. One can add to these geophysical investigations the value of palaeontological studies in providing an understanding of the development of southern floras and faunas and their dispersal, demonstrating the past links between the present southern continents.

Glaciology and oceanography

Antarctica is a unique natural laboratory for the study of the onset of continental glaciation and the formation and influence of Antarctic Bottom Water on world climate and sediments. This water mass is the main driving force of southern thermohaline circulation, and its influence extends into the northern hemisphere. The development of the present Southern Ocean circulation, and glacial history on a time scale of millions of years, can be established by studying high latitude sediments obtained by bottom coring. In addition, the bottom sediments of the Weddell, Ross and Amundsen Seas represent probably the largest unexplored body of sediment in the world, very likely with important hydrocarbon resource potential. Such sediments are well exposed and accessible at places along the eastern shore of the Antarctic Peninsula where they can be studied by land-based geologists.

It is thought that the Antarctic ice sheets and their fringing ice shelves developed about 30 million years ago. The recent climatic history of the continent can be reconstructed from studies of deep ice-cores extending back to about 150 000 years ago,[3] and there are plans for deep drilling of the ice sheet at a number of localities. Considering that the average ice sheet thickness is 2800 m (and up to 4500 m thick in places), this research is at an early stage but has great potential. Such cores can provide information on past air temperatures, accumulation rates, volcanic and solar activity, sea ice extent, and atmospheric carbon dioxide levels. Moreover, because of its remoteness, the ice sheet provides a record of background levels of atmospheric composition and global pollution which is immeasurably more useful than records from the northern hemisphere.[4] Meteorites and cosmic dust are concentrated in a few locations by a 'conveyor-belt' mechanism of ice-sheet flow, and, as unique representatives of materials from the cosmos, with terrestrial ages of up to 600 000 years, are of great interest to cosmologists.

A major concern for mankind is the stability of the Antarctic ice sheets; if they were to melt completely, sea level would rise by some 60 m, and even a relatively small acceleration of the present rate of melting, producing a sea level rise of, say, four or five metres, would have serious consequences for many of the world's cities. This is one reason why studies of the mass balance of ice sheets, and of the interactions between ice shelves, ice sheets and the ocean are important. These problems can be investigated by techniques which involve the deployment of sensitive tilt meters in order to determine the position of the grounding line of ice shelves, even through 2000 m of overlying ice. Monitoring any changes in grounding line position over the years can establish trends for predictions of future ice mass balance. With the deployment of polar-orbiting satellites, it is also now possible to measure accurately the total volume of Antarctic ice by integrating the results of satellite altimetry, which measures the surface contours of the ice sheet, with extensive radio echo-sounding measurements of ice thickness. Estimates made in this way at regular time intervals, say every 10–20 years, would have an extremely important predictive value.

The Antarctic is a major heat sink in the world climate system, and the ebb and flux of the surrounding zone of pack ice also has a very significant influence on the earth's albedo (and therefore climate), by enlarging the most effective reflecting area, and so changing the earth's heat balance. Studies on the formation, consolidation, break-up and melting of sea ice are therefore important in global terms. As well as the effect on albedo, the variation in sea-ice extent affects moisture and heat exchange between ocean and atmosphere; 10 per cent of the world ocean is south of the Antarctic Convergence. It is also a major factor in the carbon dioxide cycle; it is estimated that some 30 per cent of carbon dioxide emitted into the atmosphere is taken up by the Southern Ocean, and this has importance particularly in relation to the postulated 'greenhouse effect' which is currently a cause for concern. Experiments are planned to study ice melt-rate, radiation balance, freezing processes, water circulation, and moisture and gas exchange.

In fluid oceanography there is great scope for more studies and for the modelling of processes, such as at oceanic fronts, in relation to the oceanic gyres occurring around Antarctica, the major currents such as the Antarctic Circumpolar Current, and above all the coupling of the Southern Ocean with the World Ocean.

Meteorology and climatology

Climatologists are concerned with the inherent variability and possible man-made perturbations of the earth's atmosphere. Their raw data are the time series of meteorological records (weather- and climate-related measurements which as yet extend over only a few decades), and the much longer-term records in the ice sheets and deep sea sediments.

The ultimate source of energy for the circulation of the atmosphere is solar radiation, which is of lower intensity near the Poles. The earth's surface is heated directly, and on a lesser scale so is the atmosphere, with various processes transferring energy between the two. As mentioned earlier, the ice sheet, a high cold dome, and the surrounding pack ice are strong reflectors of incoming radiation, so that Antarctica has a large radiation deficit, which in turn has a major effect on the global circulation.[5]

After the autumn (March) equinox, the atmosphere of the southern hemisphere cools, and a winter stratospheric vortex forms due to the temperature gradient between high and low latitudes. The westerly winds create a vast whirlpool of air, with a cold dense core over the Pole, which extends even as far as the tropics at an altitude of 40 km. Unlike the situation in the northern hemisphere, this vortex persists relatively undisturbed until well after the spring (October) equinox. Associated with this, the stratosphere subsides and amounts of ozone increase, which in turn leads to the summer circulation, characterised by higher temperatures, anticyclonic easterly flow, decreasing amounts of ozone, and the disappearance of the strong polar vortex.

Whilst understanding the general circulation and inter-hemispheric differences is important, studies of the Antarctic stratosphere also have immediate practical applications. Stratospheric ozone shields life on earth from the adverse effects of ultraviolet radiation, and there is some topical concern about a possible decrease in ozone, due especially to the photodissociation of chlorofluorocarbons (chemicals widely used as aerosol propellants and refrigerants) which releases chlorine which in turn reacts with the ozone. Long-term research in the clear Antarctic atmosphere indicated no significant trends in the ozone content up to 1973, but a recent study[6] has shown that the spring values of total ozone above the Antarctic have since fallen dramatically, although the circulation has shown little change. The cause of this phenomenon is not clear, but the hypothesis is that it is directly related to the release of inorganic chlorine by the photodissociation of chlorofluorocarbons in the stratosphere. The findings of Antarctic research have now been conclusively confirmed by re-examination of satellite data, and are of considerable concern to those seeking to limit chlorofluorocarbon manufacture and use.

Related measurements, for example of water vapour and carbon dioxide in the troposphere and of ocean surface temperatures, are being made from ships and satellites. These studies are aimed at determining chemical 'sinks' (particularly in the Southern Ocean) and transport between the hemispheres. They will improve assessments of changes in climate due to the 'greenhouse effect' – the build-up of gases introduced into the atmosphere by man's activities that could cause an increase in temperature and a possible melting of the ice sheets, with the consequences already mentioned.

International collaboration in this area is good; the Middle Atmosphere Programme (MAP), concerned with the altitude range 10–120 km, has an important polar component, as has the World Climate Research Programme (WCRP). MAP is concerned with the energetics, dynamics and composition of the different atmospheric layers in this region, and the WCRP aims to improve understanding of climatic change in an attempt to develop some predictive capability.

Regular surface and balloon sonde (upper air) data are important for weather forecasting as well as for climatic research. Whilst the network of Antarctic stations is still very sparse, and concentrated particularly around the Peninsula region, the development and deployment of unmanned weather stations in the next few years should increase the overall density of observations. Antarctic data are already being used operationally for global weather forecasting, and satellites supplement the ground-based observations by, for example, remotely sensing developing cloud systems.

In addition, specific meteorological experiments are being planned to improve our understanding of physical processes. During the 1986–87 season the International Weddell Sea Project will cover aspects of meteorological, oceanographical, and sea ice investigations. Buoys deployed in remote locations will report meteorological data via satellite, and will be supplemented by data from remote sensing satellites.

Satellite imagery allows the study of the earth's surface from space, giving information on rock outcrops, shelf ice and sea ice distribution, among other aspects. Microwave imaging shows the latter, and has demonstrated the existence in some years of a previously unknown major oceanographic anomaly, the Weddell Polynya.

The ionised atmosphere and geospace

Antarctic research on solar-terrestrial physics is concerned with the ionosphere and the magnetosphere. About 99 per cent of all matter in the universe is in the form of plasma, an electrically charged gas which also forms the upper atmosphere above some 70 km altitude. The plasma in a series of layers up to about 300 km is called the ionosphere; the magnetosphere is a volume of space in the vicinity of the earth, dominated by the earth's magnetic field, which determines the physical behaviour of the plasma. The magnetic field lines connecting the northern and southern hemispheres, although extending far into space, remain within the magnetosphere, which also shields the earth from the solar wind. This gusty wind is a supersonic stream of electrically charged particles coming from the sun. It compresses the earth's magnetic field on the sunward or day side, but on the night side (beyond about four earth radii) it is drawn out into a tail almost a million kilometres long. A visible effect is obvious on occasions when solar wind plasma penetrates the magnetosheath and mixes with plasma nearer the earth, producing the brightening of the auroral displays at 100 km altitude. Above the ionosphere is a relatively cool and dense plasma, the plasmasphere, that co-rotates with the earth, separated from the warmer and thinner space outside by a sharp boundary, the plasmapause. The ionosphere and magnetosphere, taken together, are now termed geospace.

The southern polar regions are better for studies of these regions and processes than the north because there are permanent scientific stations in the Antarctic, whereas the Arctic is centred on an ocean.[7,8] In addition, the south magnetic pole is displaced further from the south geographic pole than is the corresponding case in the northern hemisphere, and this large asymmetry means that there are greater diurnal and seasonal variations in the south, as well as more varied combinations of magnetic dip angle, field intensity, and geographic latitude than in the north. For example, the stations Faraday and Dumont d'Urville are at a similar geographic latitude but their magnetic dip angles differ markedly; consequently the magnetic field lines lie in the plasmasphere at Faraday, but at Dumont d'Urville they run out into the tail of the magnetosphere.

Such extreme geophysical properties can be exploited by systematic observations, with detailed studies of particularly significant events related to the level of magnetic disturbance caused by the solar wind. The extreme conditions above and near the Antarctic are particularly suitable for testing theories, especially the wind theory and electric field theory of the F-region (the layer of maximum density in the ionosphere).

A variety of techniques are used to study these phenomena; ionosondes and more advanced ionospheric sounders that employ computer-controlled radar to analyse ionospheric echoes, magnetometers, magnetic pulsation recorders, riometers, VLF (very low frequency) radio receivers, and direction-finding equipment which can reveal details of the shape of the plasmapause and the positions of the ducts in it. All-sky cameras, photometers, and low light-level television are used to study the aurora.

Balloon and rocket experiments are being undertaken to study the ionised upper atmosphere directly, and orbiting satellites can measure the properties of the upper atmosphere and space plasma as they pass through it. The major International Solar-Terrestrial Physics (ISTP) Programme plans to deploy up to six spacecraft from 1990 onwards to examine the solar wind and various parts of geospace. Together with observations from rockets, balloons and ground observatories, including some in the Antarctic, plasma processes common throughout the universe can thus be studied in detail. The Antarctic is most important for the ground-based investigations.

Biological research in the seas

The true Antarctic biota are found south of the Antarctic Convergence or Polar Front, a physical boundary at an average latitude of about 50° S where cold, northward-flowing Antarctic surface water meets warmer southward-flowing water. The circumpolar Southern Ocean is the windiest in the world and the most turbulent, and because of this the unicellular plants which are the primary producers are unable to maintain position in the sea to make optimal use of the radiant energy from the sun. Although nutrients are not limiting, levels of primary production are therefore not significantly higher than in other oceans. Despite this, the higher trophic levels – seabirds, seals and whales – are more abundant than in other oceans.[9] This paradox is a singular feature of the Antarctic. Possibly it is to be explained by different pathways of nutrient cycling and the key position of an unusually abundant and long-lived pelagic crustacean, the Antarctic krill, *Euphausia superba*.

As research progresses, ideas on the complexity and structure of Antarctic marine food webs and the rates of energy transfer are being revised.[10] The international programme Biological Investigations of Marine Antarctic Systems and Stocks (BIOMASS) has provided an unmatched basis for co-operative studies on these problems. Although due to end in 1989 it is to be hoped that the very successful international co-operation achieved under its aegis will continue.

At the pack-ice edge, the ecology of ice-associated algae, bacteria and protozoa deserves special attention. Wider studies within the pack-ice zone are also important, although logistic problems of winter work in this environment are severe.

Research on krill is important because of its central role in the ecosystem, and also because of the commercial fishery. Although the latter has not yet reached a significant level, the future of the higher trophic levels, including squid, fish, seabirds, seals and whales, is dependent on the krill stocks being maintained at adequate levels of abundance. Improved knowledge and understanding of the status of krill stocks and krill consumers is essential to the success of the Convention on the Conservation of Antarctic Marine Living Resources. There are many exciting research opportunities concerned with the behaviour, ecology, physiology and biochemistry of krill, including large-scale studies of distribution, abundance and reproduction, and the swarming behaviour and its relation to feeding. Experimental studies on swimming, feeding, oxygen uptake rates, swarming behaviour, moulting and activity patterns, metabolism and biochemistry need to be undertaken.

All other levels of the Antarctic marine ecosystem provide exceptional opportunities for the advancement of knowledge of marine systems generally. Cephalopods, especially oceanic squids, present a particular problem. We know from their occurrence in seabird, seal and sperm whale stomachs that they are abundant and ecologically important; the problem is to find more effective ways of sampling and studying their populations. A breakthrough in this area would provide very important opportunities. Fish are probably less important in the Southern Ocean than in the Arctic, and have already been greatly overfished by commercial trawlers. Scientific opportunities perhaps lie more with the study of their physiology and biochemistry, and with their adaptations to the unusual Antarctic conditions.

Antarctic seabirds have no fear of man, and the availability in several localities of large populations of known-age, banded individuals offers incomparable opportunities for the study of seabird behaviour, physiology and ecology. Antarctic research is potentially at the forefront in these fields. The rapid development and application of new methods of study (automatic nest weighing, activity recorders, time-depth recorders, radio and satellite tags, and labelled isotopes for energetics studies) is leading to an explosion in knowledge. Ecological studies of the role of seabirds as predators on Antarctic marine resources, especially krill, are of particular importance. Similar opportunities are afforded by the Antarctic seals, including population structure and dynamics, density-dependence, social organisation and behaviour, and energetics. There is also an unusual opportunity to study factors associated with the recovery of the Antarctic fur seal from near extinction to a position as the most abundant eared seal in the world, the changes in elephant seal population ecology following the end of commercial exploitation, and the response of the world's most abundant seal, the crabeater, to the presumed increased availability of krill consequent on the reduction of whale stocks. Similar opportunities to document the continuing increase in penguin populations exist, particularly in the Scotia Sea. However, with the recent decline in commercial whaling in the Antarctic, an industry expected to end in 1988, the opportunities for whale research in the region are now very limited.

Studies of the inshore marine ecosystem have been in decline in recent years. Areas of particular interest still to be scientifically exploited are the sub-ice shelf communities, far from locations of primary production, and physiological and biochemical adaptation to the extreme environment. The long-lived species accumulate pollutants and may afford an opportunity for globally important baseline studies of pollution.

Biological studies on land

The land area of the Antarctic is 14 million km^2, but no more than about 0.25 million km^2 is uncovered by ice and therefore capable of supporting higher forms of life. Inland waters cover a relatively small area, less than 5000 km^2, and many are permanently frozen.

The Antarctic provides ideal opportunities for investigating and modelling fundamental ecological processes – of biogeochemical pathways, colonisation, succession, community development and cyclical changes. There are no large herbivores, relatively few species of microbes, plants and animals compared with other regions, and a high level of endemism, so ecosystem development is simple; energetic rates tend to be cold-adapted and slower than in temperate systems.

The main habitats to be studied include the predominant fellfield, more advanced moss and vascular plant communities (including palynological profiles of peat banks in the sub-Antarctic islands and Antarctic Peninsula), and unusual habitats such as the dry valleys and geothermally active areas; life is even found in snowfields in the form of algae and microorganisms.

Research to promote understanding of the ecosystems and their component organisms includes studies of environmental variables – mechanical, chemical and biological weathering of rocks, and microclimate. Colonisation can be studied in relation to time elapsed since glacial recession by reference to moraine terraces of known age. Opportunities for biochemical studies include investigation of the seasonal changes in soluble carbohydrate and amino acid content of coloniser microorganisms, and the effect of freeze–thaw cycles on the release of carbohydrates from mosses and lichens. Wind, birds and sea transport all have a role in initial colonisation, and the potential colonists can be assessed by culturing soil. There are opportunities for formulating principles involved in the present distribution and colonisation, both on the offshore islands and the continent.

Survival depends on adaptations to the environment, and there are distinctive characteristics relevant to the Antarctic biota. Biochemical adaptations to desiccation and freezing and, at higher levels, field activity, energetics and physiology all need to be studied. The principles involved can be related to other more complicated systems.

The relatively species-poor ecosystems, and the minimal interference with them by man, present conditions unusual elsewhere in the world, and facilitate the testing of hypotheses about biological strategies. They offer the possibility of studying the dynamics of food webs and constructing meaningful functional process models which may profitably be compared with the situation in, for example, species-poor hot deserts. The very low levels of man-made pollutants have important advantages for microbiological studies. The extremely low density human population and the lack of any significant land use by man makes it possible to dedicate land to monitoring sites for long-term studies of plant and animal communities.

The inland water bodies are a complementary part of the terrestrial system, influenced by events and processes in their catchment areas. These lakes have the added advantage that they are naturally closed systems, particularly when sealed by the formation of surface ice. They have even simpler ecosystems than the land. Like the terrestrial communities they are affected by fewer external influences than lakes in lower latitudes, and may be expected to reveal basic ecological principles more readily.

Human biology

Antarctic communities are often isolated for long periods of time, and many are small and provide 'captive' subjects of a relatively limited age range. This can be utilised for investigations into metabolism, nutrition, physiology, behaviour, psychology, and virology (notably common cold research). The effects of isolation on behaviour can be studied over long periods.

The communities also experience rigorous environmental conditions, both when working around some of the high latitude stations in winter, and on scientific field work, when even summer conditions can be extreme. The most extreme condition of prolonged severe cold plus anoxia due to altitude is most unusual elsewhere in the world. Diving in polar waters is another extreme condition, often associated with a rapid drop in skin temperature and a more gradual fall in core temperature; both may affect the diver's judgement and safety. Other extreme conditions, for example the day length regimes, offer opportunities to investigate behaviour, circadian rhythms, and endocrine physiology related to changes in day length.

These situations provide opportunities for research not just of intrinsic value, but with applications (such as the effect of cold on performance, the relation of accidents to cold, and the effect of light and darkness on performance), of relevance to other situations – for example to hill walkers and mountaineers, fishermen, commercial divers, offshore oil platforms and other isolated populations. The results of such research will also be very valuable if and when commercial or industrial activities are introduced to the Antarctic.

Arrangements for continued scientific research in Antarctica

Why should scientific research be carried out in this remote, supremely hostile and hazardous environment? First, research is a necessary qualification for achieving consultative status under the Antarctic Treaty. Second, intellectual curiosity is a driving force in science generally and Antarctica constitutes a tenth of the earth's surface and is an ideal place from which to study the upper atmosphere and geospace. Third, there are vast renewable resources of interest to man – krill, fish and squid – and research provides a basis for their rational management and conservation. Non-renewable resources are probably also present, although there are no deposits known which it would currently be economical to exploit. Fourth, Antarctica offers ideal opportunities for studying global environmental problems, including sea-level change, weather patterns and climate, and global levels of atmospheric constituents such as ozone.

The foregoing sections have encompassed a very broad spectrum of research, ranging from below the sea-floor out into space. Despite this diversity there are two unifying themes to Antarctic research: the relative simplicity of the problems, and their uniqueness. Although much of it, as with all research, is pedestrian, it is also true that there are problems being tackled which are of global importance, and of great potential or direct relevance to man. The position of science in the Antarctic is paralleled by the political arrangements, which make this 'a continent for science'. Because the Antarctic is internationalised without national boundaries or EEZs, international scientific collaboration is simplified and its extent is incomparable. This is made possible by the absence of any significant human population in the region, and by the present lack of industrial activity, other than offshore fishing for krill, fish and whales.

The factor to be balanced against these advantages is of course the remoteness of the Antarctic and the rigour of the environment. This makes Antarctic research expensive, but increasing numbers of nations are setting up Antarctic research programmes. The content of these programmes indicates that the objective of the research is primarily basic or strategic (having a potential return in, say, 25 years' time), rather than applied. So far the applied element in research has mainly been concerned with the conservation and wise management of resources, rather than directly with exploitation. It seems likely that this will continue to be the case for several decades at least.

One must not overlook the fact that scientific research is one way to maintain a presence on the continent in a cost-effective way, apart from being the only route to consultative membership of the Antarctic Treaty and involvement in the Antarctic Treaty System. Although research in the Antarctic is expensive, it is certainly cheaper than maintaining a military presence; for example, in 1982 the annual cost of maintaining a single small British naval ship in the area, HMS *Endurance*, was roughly equivalent to the total cost of the British Antarctic Survey.

Article III of the Antarctic Treaty provides that scientific information be freely exchanged, and this is usually the case in practice, while allowing some latitude for the discoverer to cream off the best results and to establish priority before making the data freely available. Also, in the case of extensive data archives there may be practical difficulties in copying the data which would penalise the originator if immediate availability was mandatory. Clearly some delay should be permitted, and it is generally accepted that such a delay, of up to two or three years, is quite reasonable.

More serious potential problems exist where the data are not purely of scientific value, but have some potential economic value; this militates against the free exchange of information. Although this has rarely been put to the test (because procrastination can postpone realisation that there is no intention of making data freely available), there is at least one field where scientific advances and management applications have been frustrated; a few countries have refused to release data on the distribution of their fishing fleets and details of krill and fish catches. As a result, the work of the Scientific Committee of CCAMLR is being frustrated. Another area where the principle of full exchange of data has probably not yet been tested concerns marine seismic geophysical data, which are relevant to surveys of sedimentary basins with potential value in hydrocarbon exploration. The ATS needs to give its attention to the problem of distinguishing between pure or academic research activities and scientific data, and commercially directed investigations and their results, as well as to clarifying and resolving the problems arising. Such difficulties are not of course unique to the Antarctic.

The initiative for collaboration in scientific research comes almost entirely from the scientists, and the vehicle for the exchange of ideas and planning is the Scientific Committee on Antarctic Research, part of the non-governmental International Council of Scientific Unions (ICSU). The Treaty System coupled with the arrangements under SCAR provides an umbrella for scientific research which is unparalleled anywhere else. The consensus among scientists is that it is successful in practice, although it is irritating when a Treaty meeting calls for advice, which involves considerable effort (and diversion from mainstream science), and on receiving a substantial report treats it in a cavalier fashion. But this is perhaps to be expected; the Treaty meetings are attended by diplomats, international lawyers and other officials, most of whom have no under-standing of scientific methods and objectives, and whose priorities are determined by political expediency.

Expansion of the Antarctic Treaty System is likely to put greater strains on the scientific community for two reasons. First, there are signs of this already in the operation of CCAMLR, where the Scientific Committee commits scientists' time but is inevitably a political, not a scientific forum. The number of subsidiary workshops and other meetings add to this concern, with progress in tackling the real problems frustratingly slow. A critical philosophical issue yet to be addressed is where the burden of proof should be – with the exploiters to show that their activities have limited adverse consequences, or with the conservers to prove the converse. (There is the alarming history of the decline of the International Whaling Commission as a disturbing indication of how it might go.) This is in marked contrast to the effectiveness (in terms of science) of the SCAR Working Groups and Groups of Specialists. The problem is that in international conventions such as CCAMLR the formal provision for a scientific committee to address scientific problems, and so far as possible give unbiased scientific advice to the civil servants and lawyers, is in practice not working. This is because overt political considerations are introduced into what should be a scientific forum, and some (if not all) scientists are constrained by political concerns, and even by instructions in their briefs. This reduces their freedom of action and compromises the outcome, thus giving critics of the system justifiable cause for complaint. Scientific principles and conclusions cannot be objectively established by politically motivated speeches. This has brought CCAMLR in particular under strong attack from the non-governmental conservation organisa-tions, and many environmentalists believe that the same will apply to minerals issues and decisions. The situation in the ATS can only get worse as scientific questions on mineral resources arise, and as the world conservation movement, often misguided, gathers force.

Second, the countries which were first to join the Antarctic Treaty are in general those with the strongest scientific backgrounds and motives. A proportion of the newcomers probably have less to contribute and may even have a hampering effect on the international scientific programmes, which point the way ahead. Some of these international programmes have been mentioned, including the very successful BIOMASS (an international terrestrial biological programme, BIOTAS, is now being planned). In the earth sciences, international collaboration has been more limited, but a major breakthrough in understanding the relationship between Greater and Lesser Antarctica may come from a major, and very costly, geoscience traverse at longitude 90° W requiring large-scale international collaboration, and which geophysicists and geologists have begun to plan. It is doubtful whether any single nation could mount such an operation. Ultimately, in terms of mineral exploration, the interpretation of geophysical results requires the drilling of deep stratigraphic test wells, and this will also depend on the pooling of resources internationally. Initially planned around refraction seismic soundings, the more informative reflection seismic techniques have yet to be employed through the ice sheet but will be needed in the future. Similarly, an adequate network of radio echo-sounding lines and satellite radar altimetry coverage, to determine ice depths to bedrock and to estimate the total volume of ice (as described earlier), will require international collaboration on a scale much larger than any hitherto attempted. Advances in atmospheric geophysics will come from the deployment of sophisticated and expensive experiments which will also involve the sharing of resources and experience. The ISTP, referred to above, is an extreme example of the vision and resources that will be deployed in this field of science.

Conclusions

In recent years the Antarctic Treaty System has been accused of being an exclusive club of developed nations. Suggestions for alternative regimes have been based either on the more or less pristine nature of the Antarctic environment (for example the 'World Park' concept), or on the argument that Antarctic resources are the common heritage of mankind (with parallels being drawn with the Law of the Sea Convention and the Outer Space Convention). The smaller nations, led by Malaysia, are concerned to take power from the hands of the exclusive Antarctic 'club', and have called for a fully international regime under the United Nations, so that the 'spoils' are widely, and possibly fairly, divided.

There is a real danger that none of these proposals would be to the advantage of scientific research. They might well diminish the willingness of governments to fund science if the present very cost-effective arrangements evolved in parallel between the ATS and SCAR were replaced by cumbersome international bureaucracies, red tape and paperwork. Added to the restraints of environment, logistics and finance would be the bureaucratic ones of permits and reporting. Consequently, the lead time for planning and conducting research would probably be spun out by protracted negotiations, and scientific effort would have to be diverted into unproductive activities, perceived as unnecessary by the scientific community. Some talented, innovative scientists who need the flexibility to respond rapidly to advances in their field of study would turn their attention to more scientifically rewarding and less frustrating endeavours. Antarctic science and the world would lose out.

5

Living resources and conservation

There are two potential conservation threats posed by activities in the Antarctic: the over-exploitation of living resources, and the impact on the ecological systems of human activities such as the logistic support of scientific expeditions, tourism and mineral exploitation. These two aspects are explored in turn below. In each case the problems of reconciling conservation and development within Antarctica are complicated by the fact that the ownership of the area is either in dispute or in doubt.

Exploitation of living resources

In the case of direct exploitation of Antarctic living resources we are concerned only with the marine environment. This is particularly a problem since the Southern Ocean is one of the few areas where productive continental shelf waters do not lie within an operating 200-mile EEZ (the exception within the Southern Ocean being around the Iles de Kerguelen, where the French authorities have imposed one). It is thus comparable to most of the waters over continental shelves in the pre-UNCLOS era. Much of the over-exploitation of the world's fish resources that occurred in the years before UNCLOS was directly attributable to this lack of ownership. Hardin's 'tragedy of the commons'[1] has a direct parallel in fisheries where, in the absence of control, new investment is attracted into a fishery until profitability is entirely dissipated and the resource depleted.

Historically, this open access nature of the living marine resources of the Antarctic has been instrumental in leading to the depletion of different groups. This process started with seals in the eighteenth and nineteenth centuries, moved on to whales in the twentieth century, and in recent decades the fin fish have become the main target of gross over-exploitation. With the exception of the whales, which until recently continued to be depleted under the management of the International Whaling Commission, all the depletion of fish and seals occurred prior to any international management agreement.

The current situation is that there are three conventions that cover the marine living resources in the Antarctic: the International Whaling Convention (IWC), the Convention for the Conservation of Antarctic Seals (CCAS), and the Convention on the Conservation of Antarctic Marine Living Resources (CCAMLR). Although these conventions are not formally part of the Antarctic Treaty, they are universally recognised as having appropriate jurisdiction within the Antarctic area. The question is whether they are adequate to ensure the conservation of the resources they are designed to protect.

The current situation

It is probably fair to say at the outset that current conservation problems of direct exploitation are small in scale. The species concerned are discussed below, in the order in which they have been exploited.

The stock of fur seals, which had been reduced to a few hundred individuals at the start of this century, has staged a dramatic recovery and currently numbers in excess of a million individuals. There is no direct commercial harvest for any seal species in the Antarctic, and even if one were to begin (which seems unlikely given current attitudes to the killing of seals) the interim catch limits under CCAS are small compared to the estimated size of the populations. Accordingly, as long as CCAS were to be observed conservation problems would be slight.

The larger baleen whales have been protected by various decisions of the IWC and may be expected to recover slowly, although there is little direct evidence of such recovery to date. The minke whale is currently harvested by Japanese and Soviet fleets, who operate under the objection procedure of the IWC. Whether further diplomatic pressure from countries who support the cessation of all whaling will achieve a halt in the remaining Antarctic whaling is open to question. However, current harvest levels are small compared to the estimated stock size, and, barring gross scientific errors, dramatic depletion of the minke whale is unlikely, at least in the short term.

Fin fish are currently harvested under the auspices of CCAMLR. Some conservation measures and resolutions have been brought in, the most important in recent years concerning the marbled notothenia (*Notothenia rossii*). Directed fishing for this, the most depleted of the fin fish, is prohibited in all areas of the Southern Ocean. In addition, various resolutions have been adopted which are concerned with ensuring that the by-catch of this species, which occurs in the directed fishing for other species, is minimised. However, the main stocks of *N. rossii* were depleted some fifteen years ago, and current levels of abundance are estimated to be less than a tenth of the unexploited level. Accordingly, the measures that have been adopted are irrelevant to the industry, and the status of the remaining species of fin fish is largely unknown (with the exception of around the Iles de Kerguelen, where the French authorities have been monitoring the stocks). CCAMLR has singularly failed to institute precautionary measures in the light of this ignorance. In global terms, the fin fish of the Antarctic are of minor importance and further depletion is unlikely to be a particularly attractive economic prospect. Nevertheless, the appropriate management of fin fish is currently the main conservation problem in the field of direct exploitation.

Krill catches reached a peak of over 500 000 tonnes in the 1981–82 season, and have since declined to under 200 000 tonnes. Since estimates of the potential yield of krill are of the order of tens of millions of tonnes, such a harvest is negligible.

One of the central concerns of those involved with drafting CCAMLR was not the direct over-exploitation of krill itself, but rather that the level of krill harvesting should be kept within bounds that ensured that species dependent on krill for their food supply were not threatened by an erosion of their food resources. This is a real and proper concern. The baleen whales, several species of seals and numerous species of birds depend on krill for their livelihood, and it is quite conceivable that large-scale krill harvesting could adversely affect these species. There are a number of scientific grounds for believing that the most likely impact would be on shore-based species during the breeding season. Successful foraging for food to succour young is dependent on the density of krill in the vicinity of the bird and seal colonies.

The operation of fishing fleets in an area in direct competition with foraging adult birds and seals is therefore the most likely scenario for conservation problems. Baleen whales, by contrast, are relatively free to move large distances in search of their food, and are therefore less vulnerable to local depletion of krill, which is in any case unevenly distributed throughout the Southern Ocean. Only if a krill fishery were to increase to a level where millions of tonnes were removed, would it be likely that the baleen whales would suffer. It is clear that current harvest levels are unlikely to present significant conservation problems.

The future
That CCAMLR is already in place before large-scale krill exploitation has started, that the CCAS has interim catch levels of a conservative

kind, that the IWC has recommended a pause in whaling, and that current levels of harvest are small for the majority of species are all grounds for optimism. However, there are still grounds for concern about the future.

At the moment there is very little economic pressure to harvest beyond sustainable levels (with the exception of the fin fish), but the history of fisheries abounds with cases of dramatic expansion when economic possibilities become attractive. The sort of situation in which krill harvesting might increase dramatically has been discussed in a study by the International Institute for Environment and Development.[2] It may be assumed that in such a situation the economic pressure for unrestricted expansion would be considerable. Similar considerations would apply to the harvesting of seals or other species. It should be noted that in 1987 there was exploratory fishing for squid around South Georgia following the restrictions placed on that industry in the area around the Falkland Islands. Were there to be great pressure for rapid expansion of any of the above industries, there must be doubts as to whether CCAMLR, CCAS or any multilateral regime would be successful.

International Fishery Commissions have rarely been successful in implementing effective management measures. In many bodies the recommendations for management measures had to be by consensus, and if they could be approved by a majority vote then there were usually arrangements whereby a country could object to the proposal, so that it was no longer binding on that country. The Commissions were therefore slow in introducing measures, and most that were introduced, such as controls on the mesh size of nets or on the sizes of fish, put no great restrictions on fishermen's activities. Measures that control the volume of catch or amount of fishing have been much harder to get accepted, and the progress of CCAMLR to date has been typical in this respect.

In the north-west Atlantic it was the delays in the introduction of quotas for haddock and other depleted species that influenced the United States and Canada to press more heavily for changes in the Law of the Sea as it related to the management of fish stocks. The introduction by the International Commission for North Atlantic Fisheries (ICNAF) of quotas for haddock and later for other species was largely due to the realisation that, if it failed to act along the lines clearly suggested by its scientists, then the United States and Canada would act immediately and with good justification to extend their territorial limits without waiting for any new Law of the Sea arrangement.

By contrast, in the 1960s the North-East Atlantic Fisheries Commission (NEAFC) presided over the decline of the North Sea herring stocks. Despite clear scientific advice from 1964 onwards, it was not until 1968 that some action was taken (a closure of the southern part of the North Sea, and some other parts), and that fell well short of what was needed. A near-complete collapse of the stock then followed, and it is only now recovering.

One big problem in the management of fisheries by international Commissions has been the treatment of uncertainty. Some doubts and uncertainties will exist in any scientific analysis of a fish stock, and honest scientific advice must reflect those uncertainties. Under international arrangements these doubts have been used by some interests to weaken action and to seek less restrictive measures, for example higher quotas, by focusing attention on the more optimistic of the estimates. With the need for near consensus under such international arrangements, these views tended to prevail. A single national government with jurisdiction over the resource can take a balanced view, or, if it wishes, take a deliberately conservative view, perhaps allowing only the volume of catch that the scientific evidence suggests is safe.

In this context it is instructive to consider the success of the French in managing the fin fish resources around Kerguelen, in comparison with the rather inadequate measures invoked around South Georgia and in other areas of the South Atlantic. In ignorance of the status of the stocks, the French closed the fishery until reasonable scientific estimates could be made. Within CCAMLR, despite a clear majority of the Scientific Committee advocating closure of the fishery around South Georgia, no such measure was achieved either in 1984 or 1985.

For fin fish the problems involved are relatively straightforward scientifically, and the burden of proof once sufficient data are available is not particularly onerous. By contrast, the sort of scientific problems generated by the wording of the CCAMLR are complex; CCAMLR concentrates not just on the conservation of individual species, but on the conservation and protection of the ecosystem as a whole. This means that the data needs are new (and in some cases ambiguous), and that there is no all-embracing and accepted body of theory. It is therefore hard to conceive that a reasonable consensus on even the scientific questions would easily be achieved. One of the complications is that scientific data can accumulate from two sources: those data that are collected incidentally to commercial operations, and those data collected by scientific research programmes. Normally, the assessment of fish stocks can be done using data acquired incidentally to fishing activity, but the data required for CCAMLR to be operated effectively are more complex, and would need to be collected as part of a directed scientific research programme. To some extent this was recognised by the scientific community who organised the SCOR/SCAR BIOMASS programme. However, it seems fair to say that the scientific work performed under BIOMASS falls far short of CCAMLR's requirements.

Historical experience, both in the recent past and in the development of previous fisheries commissions, leads to the view that some form of sovereign pressure is required to achieve satisfactory conservation goals. For example, an indication from the relevant authorities that the failure of CCAMLR would result in the declaration of an EEZ around South Georgia might well be the stimulus required to ensure that CCAMLR worked. Similar considerations might also apply in the case of any direct harvesting proposed for the now abundant fur seal colony on South Georgia.

There are some reasons for preferring an option in which an international arrangement was made to work by such pressure. International arrangements for the provision of scientific advice have few negative features and some positive advantages. For example, in the case of the fisheries around South Georgia, the United Kingdom and Argentina have few scientists with direct experience of the fisheries. International arrangements under CCAMLR permit a mix of scientists of all countries involved to take part in assessing the status of the stocks. Freed from overt political pressure, such arrangements should invariably be beneficial, yet they are only likely to work in a situation where the threat of a sovereign declaration ensured the compliance of other States.

CCAS and CCAMLR lie well within the embrace of the Antarctic Treaty System. The involvement of the UN General Assembly in Antarctic questions, and the dissatisfaction amongst some environmental groups with both CCAMLR and the environmental aspects of the Antarctic Treaty, are both pressures leading toward further internationalisation of the Antarctic. There is little reason to hope that such further internationalism will improve the ability to achieve conservation goals. This is the case whether such a tendency is manifested within the ATS itself, for example by the extension of the membership of CCAMLR, or by the full involvement of the United Nations.

Wider membership of the IWC has recently led to a reduction in whaling, yet whaling continues despite a majority vote for a pause. Indeed, the present reduction is only conceivable due to American pressure on various States via the Packer–Magnusson Acts, which have effectively imposed 'sovereign-like' pressure on the world's whaling fleets.[3] These Acts have the effect that, in the situation where a marine conservation treaty acceded to by the United States has been undermined by another State, the US must cease all trade or business in fisheries with that State. One respect where this has been particularly successful has been the elimination of 'pirate' whaling.

Conclusions

The current conservation problems of direct exploitation of living resources in the Antarctic are small in scale. However, the ability of the current regime to cope with any increase in economic pressure to harvest is in doubt. Political movement for further internationalisation of the regime seems likely to weaken rather than to strengthen its ability to successfully marry conservation and development. Successful operation of conservation regimes on exploited resources has only been achieved when some effective and authoritative management has been imposed, and there are few grounds for believing that the situation will alter in the future. In the past such management has only been achieved when States have operated an effective sovereignty over the management of the resources.

Conservation problems posed by other human activities

There are three other main types of activity which potentially pose conservation problems within the Antarctic: scientific research (in particular its logistic support), tourism, and mineral development. All these activities have, or will have, some kind of environmental impact on the flora and fauna of the region.

Currently the basic provisions of the Antarctic Treaty, together with the Agreed Measures for the Conservation of Antarctic Fauna and Flora, are the main legal instruments that govern activities in the area. The series of meetings of the Consultative Parties which have addressed the problem of setting up a mechanism to regulate mineral exploration and exploitation have yet to reach a conclusion. However, from all reports the main conservation concerns addressed are similar to those in the Treaty and the Agreed Measures. In essence the conservation provisions are of three types:

1. The provision of areas where all or particular activities are excluded.
2. The assessment of the likely effect of activities before they take place, and the implicit abandonment of those activities which have particularly deleterious effects on the environment.
3. The regulation under permit of various other potentially damaging activities so that damage is minimised (the Agreed Measures specifically prohibit interference with any mammal or bird without a permit; however, the guidelines for permit issue are variable and in certain circumstances obscure).

Such conservation provisions are reasonable in principle, but how effective are they likely to be in practice?

Scientific activity

Although some scientific activity involves the interference with or killing of specimens, the effects are minor given the population size of the fauna.

The main conservation concern over scientific activity is with its logistic support, and there are a number of reasons for this. Scientific activity within the Antarctic has increased substantially over the past two decades. The increasing number of scientific bases are concentrated disproportionately in the small area of the continent not permanently covered by snow and ice. Many of these ice-free areas contain vulnerable vegetation complexes with associated invertebrates, as well as freshwater lakes with their own communities of plants and animals. Coastal ice-free areas also support breeding colonies of sea birds and seals, which are vulnerable to disturbance. In addition, increased scientific activity has been associated with, and enhanced by, the construction of hard airstrips, docks and large buildings. The construction of all of these can have deleterious environmental consequences, and airstrips also have the secondary effect of generating increased human traffic. There have been a number of reviews of the environmental impacts of scientific operations in recent years,[4] and detailed review is unnecessary here. However, apart from the direct destruction of habitat needed for station construction there are other concerns over waste disposal and the activities of personnel.

The central problem lies in the development of existing bases and the siting of future stations. In theory the Treaty System provides some protection by the designation of Specially Protected Areas (SPAs) and Sites of Special Scientific Interest (SSSIs). In the past, regrettably, practice has been very different. Consider the case of the Fildes Peninsula; although originally an SPA, this site is currently occupied by five major stations, an airstrip, and several buildings. However, even if practice were to improve, areas protected by such site designation are small and cannot be considered as providing extensive protection for each individual ecosystem type.

Although in principle some form of impact assessment is required prior to significant development of a site, this is not always satisfactory in practice. In the vicinity of their own bases, individual national authorities effectively control their own activities. Accordingly, different standards will be applied when decisions on the projects are to be made. Although the recent report by an American observer team which visited a variety of bases indicates the willingness of the various authorities to co-operate in the inspection process, official mechanisms by which other States, organisations or individuals can comment on such developments do not exist.

Table 5.1 lists the activities associated with scientific work in the Antarctic whose effect SCAR believes should be assessed prior to implementation.[4] At present no mechanism exists to ensure that such guidelines are followed by the national authorities, and this must be seen as a short-coming of the current system.

Finally, there are two recent phenomena which are of concern. The first is the existence of privately sponsored scientific expeditions, whose activities clearly lie outside the normal activities of the Treaty. The second is the likely increase in demand for bases for scientific research, brought about by the membership requirements of the Treaty System. To

Table 5.1. *Categories of non-commercial activity that might reasonably be expected to have a significant impact on the Antarctic environment*

Scientific activities
1. Interference with or modification of endangered or unique systems, communities or populations.
2. Operations which might adversely affect SPAs or SSSIs.
3. Introduction of alien biota with the potential to multiply or disperse.
4. Any operation affecting areas valued mainly for their sterile or pristine nature, e.g. dry valleys, remote ice cap areas.
5. Application of biologically active substances which have the potential to spread so as to cause perceptible effects outside their area of application.
6. Operations which might perceptibly impede the recovery of any endangered, threatened or depleted populations.
7. Experiments deliberately designed to create adverse changes in populations or communities (perturbation experiments) which extend over areas of more than $100 \, \text{m}^2$ or, possibly, even less, particularly if unique systems are involved.
8. Operations which will adversely affect populations for which long time series of data have been (or are being) collected to establish the status of the population.
9. Introduction of radionuclides into the environment where their subsequent recovery and removal cannot reasonably be assured.
10. Drilling operations involving the use of drilling fluids other than water and/or possible escape or vertical movement of subterranean fluids.
11. Marine seismic surveys involving the use of explosive charges.

Logistic/support activities
1. Establishment of bases, stations, airstrips, etc, or the extension of existing facilities (by, say, more than 10%).
2. Increases in personnel or personnel movements, aircraft flights, etc (by, say, more than 10%).
3. Major changes in amount (by, say, more than 10%) or type of power generation, or fuel consumption.
4. Any operation affecting areas valued mainly for their sterile or pristine nature, e.g. dry valleys, remote ice cap areas.
5. Any operation involving the closing of a station which has been active for several years.

the extent that this enhances the demand for scarce sites in areas of vulnerable habitat, this phenomenon has the potential for environmental harm.

Tourism

Although there has been no clear trend in the number of tourists visiting Antarctica over the last decade, it seems likely that some increase will occur in the future. The vast majority of tourists to the Antarctic come via cruise ship expeditions, which have landed parties in only a few locations; for this reason their capacity for inflicting damage on the environment is limited. Such tourists have also tended to be interested in the ecology of the region, and have correspondingly been sympathetic to conservation needs. Nevertheless, the extent of control over tourism provided by the current system is almost non-existent. Tourist ships are usually registered in non-Treaty States, and control by the ATS over their activities is therefore nil.

The extent of real influence on tour operators lies in the power of the Treaty Parties to deny visits to scientific stations, and the dissemination of a tourism code of practice. Apart from such moral pressure there is little control. For example, there is nothing in the current system to stop a land-based tourist operation being set up in a vulnerable habitat, nor even to require that there be some form of assessment of the impact of such a development. Cost considerations and likely profitability at present render such a scenario unlikely, but it is an uncomfortable thought that such activities could in theory go ahead unregulated. Indeed, on the Fildes Peninsula, the Chilean station at Teniente Marsh contains a tourist hotel served by the large airstrip. This development has never been discussed by parties other than the Chilean authorities.

Mineral exploration and exploitation

It is clear that unregulated mineral exploitation in the Antarctic region could have potentially devastating consequences. Elsewhere, in the absence of environmental controls, operators have sought to maximise their profits at the expense of environmental costs. Such activities are well documented and well known. In the Antarctic, onshore mineral activity would be expected to have similar environmental impacts, with the creation of spoil heaps and other damage to the environment caused by the associated infrastructure. Since any onshore mineral activity would be likely to be concentrated in the ice-free areas, such an effect could be considerable. Offshore exploitation of hydrocarbons has the associated potential for damage by oil spills; concentrations of hydrocarbons through the food chain could also occur in principle, so magnifying the potential damage.

These and other potential impacts of mineral exploration have been examined in a number of reports including two by a SCAR Group of Specialists.[5,6,7] These reports make disturbing reading; the potential damage to a bird or seal colony by an oil spill, the take-up of hydrocarbons by krill, and the mining for shore-based minerals in the vicinity of a penguin colony are all hazards that one would seek to avoid. It is thus fortunate that the expectation of significant mineral exploration or exploitation in the next decade or so is small (see Chapter 6).

The mineral regime currently under discussion is likely to provide that certain areas be forbidden to all activity, and other activities be banned until the environmental impact of that activity may be assessed and found to be small. In principle this could be a satisfactory level of protection for conservation; however, there are some central questions to be answered. What is 'small'? Who will decide what activities are (or are not) permitted? How will such a regime be policed? And in what way can environmental impact procedures be made available to independent scientific scrutiny? All such questions revolve around the problem that was addressed in the context of the direct exploitation of living resources – namely the existence or otherwise of authoritative management.

The main problem is that in order to achieve consensus (or to avoid objection) the burden of proof will likely often be on those seeking to demonstrate that detrimental effects are likely. Yet the SCAR group indicates the substantial gaps in knowledge of the environmental impacts of a wide variety of activities. An individual authoritative management body can decide to adopt a conservative approach, but a body in which all participants must agree is unlikely to do so on all occasions. Whether such problems occur at the scientific advisory level or at the level at which decisions on the basis of scientific advice are taken is largely unimportant. In the absence of authority it is hard to envisage practical circumstances where the benefit of the doubt is given to the environment. Similar considerations apply to the problems of policing, monitoring, and ensuring independent scientific scrutiny of the results.

Conclusions

As the political process continues, the regimes which govern the conservation of Antarctica will also alter. Institutional arrangements will vary as will the membership of the institutions in an unpredictable way. Given this uncertainty, there seems to be a need for some authoritative and practical management principles which can be used to evaluate developments. Such a suggestion for an Antarctic 'Conservation Strategy' or 'Long Term Plan' has been made, and is under active development by SCAR and IUCN. The attitude of the Consultative Parties to such a mechanism is currently unknown, but it is possible that it may gain acceptance more easily in a wider international forum.

6

Mineral resources

Ever since the *Glomar Challenger* (1972–73) detected traces of methane and ethane in the Ross Sea, the oil search has been on, with even official US reports talking about 'tens of billions of barrels'. And although difficulties and costs would be astronomic – possibly twice those at Alaska's Prudhoe Bay – the searchers are not noticeably dismayed. Norway and West Germany have been looking in the East Weddell Sea, Japan has just completed a three-year trawl through Bellingshausen, Weddell and Ross, France is active off Terre Adélie, Australia has had a look at the Amery Ice Shelf, and the Soviets in the Drake Passage, with aeromagnetic surveys over the Filchner, Ronne and Amery ice shelves. Even the Poles have been echo-sounding in the Peninsula.[1]

Reading this extravagent assertion about the search [*sic*] for hydrocarbons in Antarctica, one immediately asks how this accords with the facts. Even assuming that there are hydrocarbon or other mineral resources, could they be exploited economically and in an environmentally acceptable way? To what extent could the experience of mineral exploitation in the Arctic be extrapolated to the Antarctic? These are just a few of the many questions that come to mind. This chapter attempts to assess the likelihood of mineral exploration and exploitation in the light of what is known and, more importantly, what is not known about Antarctica.

Antarctica is a continent with many unique features which have a direct bearing on any mineral exploration or exploitation. It is covered almost entirely with a permanent ice sheet up to 4.5 km thick, it is the highest of all the continents, averaging about three times the average altitude of the others, and a mere 2 per cent of the surface area is exposed rock. This last point needs emphasis, for it is still asserted by some authors that there is up to 4 per cent bare rock on the continent; this misunderstanding arises from the fact that although 3–4 per cent of the area comprises mountain ranges projecting through the ice, these mountains are themselves largely covered by ice and snow.

The ice sheet is continually flowing outward toward the surrounding seas, and as a result much of the continent is fringed by vast floating ice shelves, beyond which the sea is frozen in winter to form the pack ice which covers an area of at least 18 million km^2 (or about one and one-third times the area of the continent itself). As Thomson and Swithinbank[2] have described it, 'of the seven continents, Antarctica is the coldest, highest, driest, windiest and least accessible. With a total area of some 14 million km^2, it represents one tenth of the world's land surface. It is half as big again as the United States.'

A comparison between the Arctic and the Antarctic is revealing. The Arctic is an ice-covered ocean basin, a geographical Mediterranean; the water circulation is dominated by the Atlantic, with most of the input being from the Atlantic via the Fram Strait, circulating in the Arctic Ocean and returning to the Atlantic with only a minor contribution from the Bering Straits. The shores of this polar Mediterranean comprise the northern extremities of, in the main, the developed nations of the northern hemisphere. There is both an indigenous population (the circumpolar Inuit) and modern settlement in the Arctic, and in terms of resource exploitation any oil, gas or minerals discovered offshore or onshore can be fairly readily moved southward to the centres of population and consumption in the developed economies of the States bordering the Arctic.

In complete contrast, Antarctica is a continent entirely surrounded by sea with no indigenous population (important because it means that there has been no pressure for development from local inhabitants) and, except for the Antarctic Peninsula, separated from the potential markets for its resources by thousands of miles of ocean. These waters are made difficult and dangerous by the presence of huge icebergs (many times larger than those encountered in the Arctic), and also by the seasonal development of the vast fringe of pack ice.

Having emphasised the differences it is worth mentioning one little-known similarity, which results from the fact that much Antarctic scientific research is centred on national bases. This is the development of a series of independent enclaves, the national bases, which is reminiscent of the Arctic where, for example, the adjacent parts of Arctic Canada and Greenland have evolved quite separately. Though usually thought of as being quintessentially international in terms of scientific co-operation, the way in which the scientific settlement is evolving is more properly described as multinational.[3] A modicum of national rivalry is healthy, but to further science it must be kept in check, and international co-operation must be the order of the day. It will be essential if some of the really large-scale investigations, fundamental to understanding not just the geology but many other aspects of the physical and biological environment of Antarctica, are ever to take place.

Although in the strict sense Antarctica has no indigenous inhabitants there is a small permanent population of about 1000 over-wintering scientists and support staff, whose numbers swell to about 4000 during the field season in the austral summer. To sustain even this small body of scientists, everything – food, fuel, equipment and construction materials – has to be brought in from outside at great cost. The Antarctic research programme budget of one of the Treaty nations has a logistics/science ratio of more than 10 to 1; it does not take much imagination to see that this ratio would be vastly greater for resource exploration and exploitation in Antarctica.

Whatever scientific research or, eventually, commercial exploration, may reveal about the mineral resource potential of Antarctica, exploitation should only take place under the most stringent environmental controls; for the purpose of impact assessment it is necessary to understand the physical, chemical and biological processes at work in Antarctica. Environmental protection management principles and practices must be developed for Antarctica and be applied to existing or projected scientific bases and activities. The American McMurdo Station, for example, is probably comparable in size and facilities to what would be required for an oil exploration base, but would an application for such a base be approved if what was proposed was an exact copy of McMurdo? It is to handle such issues that the Treaty nations need not just scientific information, but more importantly, an environmental protection management system which should be operating now; the establishment of such a system is perhaps of more immediate practical concern (for example for the location, construction and operation of new bases for scientific research or the expansion of existing bases) than a minerals regime.

Antarctic geology is still in its infancy, and there is a very long way to go before exploration for mineral resources could be seriously contemplated. In order to appreciate what would be required in terms of geological knowledge before the exploitation of any Antarctic mineral resources could seriously be contemplated, it is worth considering the history of the discovery and development of the North Sea hydrocarbon province. In 1815, William Smith, often referred to as 'the father of English geology', published the fifteen sheets of his 'Geological Map of England and Wales with part of Scotland', and at about the same time Brongniart and Cuvier published the first geological map of the distribution and structural relations of the major rock units of the Paris Basin. From then on, our knowledge of the land areas of Europe grew steadily, and for the hundred years after publication of these first surveys (up to the 1920s) was based primarily on surface mapping, supplemented by subsurface information from mining, tunnelling and water-well drilling. In the 1930s, serious oil exploration began in Austria, Germany, the Netherlands and the United Kingdom, and by 1939 small oilfields had been discovered in all these countries. The results of this exploration, together with data from the emerging science of geophysics, further contributed to our knowledge of the geology of Europe, and then, in 1959, a huge onshore gasfield was discovered at Groningen in the Netherlands.

The producing formation at Groningen is a Lower Permian Sandstone capped by evaporites (including salt), which was known to be very similar to the geological succession at this stratigraphic level in the East Midlands, eastern and north-eastern England. The evidence suggested that the sediments were probably laid down in the same depositional basin, part of which now lay between the two countries beneath the southern North Sea; if this basin contained structures large enough to warrant offshore exploration there was a good probability that they would have reservoir sands and cap-rock conditions similar to those obtaining in the Netherlands. On the basis of this hypothesis a group of companies conducted a reconnaissance seismic reflection survey in 1962 to look for structures, and this was followed in 1963 by an aeromagnetic survey of most of the North Sea south of 58° N, together with some additional seismic work. The results were encouraging, but no drilling could take place because the States bordering the North Sea had not agreed the ownership of the continental shelf outside their territorial waters.

Without going into the legal details, suffice to say that the governments of the riparian States enacted legislation under the terms of the 1958 Geneva Convention on the Law of the Sea, and in 1964 settled the subdivision of the North Sea sector of the European continental shelf (either on the basis of median lines equidistant from their respective coasts or by mutual agreement). Drilling started in 1965 and the first offshore gasfield was discovered that year in the British sector, and four years later the first offshore oilfield was discovered in the Norwegian sector.

This very abbreviated account illustrates the way in which the accumulation of geological knowledge over 150 years underpinned the discovery of a major new oil province right on the consumers' doorstep. It also demonstrates the essential integration of a vast corpus of geological information with the data from geophysical surveys (and make no mistake, the former is a prerequisite for the interpretation of the latter). Finally, the North Sea experience demonstrates the necessity of a clear title to a lease or licence as a prerequisite for commercial exploratory drilling. It is interesting to speculate whether the North Sea hydrocarbon province could have been discovered and developed in the time that has elapsed since William Smith published the first geological maps if Europe had been covered by an ice sheet and the North Sea was under a floating ice shelf hundreds of feet thick. A trifle far-fetched perhaps, but it does serve to emphasise how much we need to know of geology before we can discover and develop natural resources. In Antarctica we are a very long way from knowing enough, and notwithstanding the advent of remote sensing and the sophistication of modern geophysical methods, it will take a very long time and be very expensive – possibly even prohibitively so – to reach that stage.

Antarctic geology: an overview

A brief resumé of the gross features of the geology of the continent is in order before going on to discuss the mineral prospects, exploration and how any discoveries might be exploited. The literature on Antarctic geology is growing apace,[2,4,5,6] notwithstanding the fact that much of the continent has only been mapped at not much more than a reconnaissance level. What follows is not a detailed account of the geology of the continent, but rather an outline of the salient features of the geology as a background to the discussion of mineral resources.

The history of the geological study of Antarctica dates back over 80 years to the turn of the century; there were geologists on the expeditions of the heroic era of Antarctic exploration, and men such as Mawson and Debenham made pioneering contributions to Antarctic geology. Sporadic work, largely expedition-related, continued between the wars, but it was only after the Second World War that significant advances in geology started to take place. Bases were established in the 1950s, and systematic scientific study dates from the International Geophysical Year of 1957–58; as the name suggests, the emphasis of the IGY was mainly on geophysics, but it led to an almost quantum leap in all branches of knowledge about Antarctica.

The revival of the concept of continental drift in the 1960s, as evidence of large-scale horizontal movements of the earth's crust mounted, led to the theory of plate tectonics. This theory has transformed geology and profoundly influenced the study of the continents and ocean basins. It gave new stimulus to the geological study of Antarctica, which occupied a key position at the centre of the ancient southern supercontinent of Gondwana before its break-up.

For the geologist Antarctica is tantalising; most of the continent is covered by ice and a large proportion of the continental shelf is overlain by permanent floating ice, beyond which there are huge icebergs and a seasonal cover of pack ice. On land, outcrops are limited and unevenly distributed, with no exposures at all over 98 per cent of the continent. In consequence, geophysical methods of sensing through the ice have assumed great importance, not only for determining the form of the land surface beneath, but more particularly for investigating bedrock geology. However, geophysical surveys alone cannot provide a definitive geological picture of the nature, structure and age of the rocks. In order to understand the geology, the geophysical and geological data must be integrated, and the difficulty of doing this on a continent almost completely covered by ice, with widely scattered outcrops and no evidence of the nature and age of the geological succession from boreholes, is obvious. In spite of this there has been much progress towards an understanding of the gross features of the geology of this vast continent, but even though a few areas have been studied in some detail there is an enormous amount still to be done.

Although Antarctica is often thought of as a single land mass concealed beneath an ice sheet, it is now known from geophysical studies that if the ice were to be removed it would no longer be a single continent. The sub-ice topography revealed by measurements through the ice comprises a continental landmass, Greater Antarctica, and a chain of islands – Lesser Antarctica – separated from Greater Antarctica by a deep submarine trough which extends from the Weddell Sea to the Ross Sea.

Geologically, Greater Antarctica is a stable continental block consisting of Precambrian rocks, locally overlain by relatively undisturbed sediments ranging in age from Cambrian to Jurassic. These rocks, which include Lower Palaeozoic sediments deposited in shallow marine environments, are mainly seen at outcrops in the Transantarctic Mountains, one of the world's longest mountain chains, which mark the western edge of Greater Antarctica. Much of Greater Antarctica has similar geology to India, Africa and Australia and, in plate tectonic terms, is a classic divergent continental margin linked to the separation of Africa, India and Australia from Antarctica.

Lesser Antarctica is an archipelago buried beneath the ice, comprising a structurally complex area of igneous (both volcanic and plutonic),

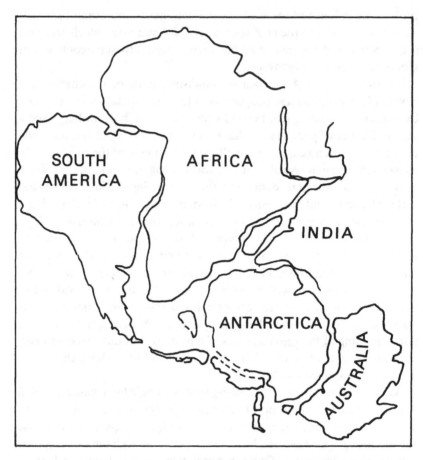

Fig. 6.1. Reconstruction of Gondwana supercontinent *ca*. 200 million years ago, showing Antarctica occupying a key position (after Thomson & Swithibank, 1985[2]).

sedimentary and metamorphic rocks, along a continuation of the Andean mountain chain. The geological evolution and tectonic complexity of Lesser Antarctica are characteristic of an active continental margin, and it was in fact the active margin of the Gondwana supercontinent facing the ancient Pacific Ocean. Behind this volcanic arc (i.e. between Greater and Lesser Antarctica) are a series of submerged sedimentary basins and troughs which extend from the Weddell Sea to the Ross Sea. From this it is clear that Antarctica is not as straightforward as it seems at first sight; if all the ice were to melt, then topographically it would appear as a large eastern continent, separated by a more or less continuous depression (occupied by a shelf sea) from an offshore island arc or archipelago.

If we now go back in time the scene changes even more dramatically. Two hundred million years ago, Antarctica was linked with India and the other continents of the southern hemisphere to form the single supercontinent of Gondwana (Fig. 6.1). Antarctica was at the nucleus of this continent, which began to break up about 180 million years ago (in the Jurassic), and the process of drifting apart has continued ever since. First South America separated from Africa, then India began to drift northward, with Australia the last to go, breaking off from Antarctica about 50 million years ago. These continents fit together well both geologically and topographically in the now almost universally accepted reconstructions of Gondwana.

The tectonic complexity of the Antarctica Peninsula and Lesser Antarctica as presently known has already been noted, and this area stands out as an exception when Gondwana is 'reassembled'. However one tries to fit the continents together, the Antarctic Peninsula overlaps South America, which is patently a geological nonsense. In an effort to resolve the problem, geologists are now investigating the hypothesis that Lesser Antarctica consists of a number of discrete crustal fragments, or microplates, which have moved separately and, in some cases, may also have rotated. It would be very difficult to prove this theory in a well-exposed, easily accessible area of the world, with good data from geological and geophysical surveys backed up by information from deep boreholes, but given the unique conditions obtaining in Antarctica it will take a very long time, be enormously costly, and (despite the optimism of some geologists) may ultimately prove impossible.

'Hard-rock' minerals

What, then, do we know of the mineral resources, actual or potential, of Antarctica? The short answer is very little. To put the question into context it is important to understand that to date there have been no specific exploration programmes aimed at finding hydrocarbons or minerals in Antarctica. Our present knowledge of the mineral resources is, without exception, based on data collected as part of a continuing programme of geological and geophysical research, to which the scientists of many nations have contributed. Not surprisingly, these field studies have sometimes recorded the presence of metalliferous and other economically important minerals and, where outcrop extent and quality permits, there may be sufficient evidence from the field relations, mode of occurrence, structural position, and subsequent laboratory studies of the samples, to get a good idea of the processes involved in the observed mineralisation. Such mineral occurrences are interesting and important as evidence contributing to the study of Antarctic geology.

One swallow does not make a summer, and one mineral occurrence does not make a deposit. The distinction is very important; in Antarctica there are many known mineral occurrences, but so far very few deposits have been identified. The following are a few examples of some of the metal ores identified at outcrop or in samples obtained during the course of geological surveys: in the Antarctic Peninsula, copper, molybdenum, gold, silver, nickel, cobalt and chromium; in the Transantarctic Mountains, copper, lead, zinc and gold; in Greater Antarctica, copper, molybdenum, tin, manganese, titanium and uranium; and in the Dufek Massif (Pensacola Mountains), chromium, copper, nickel and platinum.

Of all these, on the evidence available, the Dufek Massif would, without doubt, be the prime target if mineral exploitation were ever to be undertaken in Antarctica. The Dufek Massif will be discussed in more detail later, suffice for the present to note that it is still only a potential deposit which would have to be extensively explored to determine what it contains and to see whether the target minerals are rich enough to warrant exploitation.

To date the only known mineral deposit of any size in the Antarctic is the low-grade iron formation in the Prince Charles Mountains, and that is unlikely to be of any commercial interest until all the more accessible known iron ore bodies in the world are approaching exhaustion. There is also coal in the Transantarctic Mountains, where the Permian coal measures contain high-rank bituminous coals in seams which average 0.5–1.0 m thick. It is fair to say that if these coals were located on any of the other six continents they would compare so badly with the known deposits that they could not be economically exploited in the foreseeable future (for which a figure of about 100 years would not seem out of the question).

It can therefore be said with confidence that, contrary to the impression created by the quotation at the start of this chapter, all the evidence is against imminent commercial exploration, let alone exploitation, of Antarctic mineral resources.

The Dufek Massif

The Dufek complex, which features in every account of the potentially exploitable minerals of Antarctica, is a layered basic igneous intrusion discovered during the IGY, which has been compared with the Bushveld complex in South Africa (the source of 85 per cent of the world's platinum). Aeromagnetic surveys and radio echo-sounding suggest that it has a minimum area of 50 000 km^2, whereas the Bushveld has a known area of 67 000 km^2. Although the two are not grossly dissimilar in area it is important to note that the size and thickness of the Bushveld have been accurately delineated, whereas only a few per cent of the estimated total area of the Dufek is exposed – the rest is hidden under ice – and even the exposed area has not been 'explored' let alone 'prospected' in mining terms.[7]

Comparison of the thickness of the two bodies is less reliable, and the mineralogical comparison is hypothetical. To date only magnetite bands a metre or more thick have been mapped in the Dufek, and the chromite and platinum of the Bushveld have not been seen there. If the two are strictly comparable it is possible that there could be chromite in the deeper, unexposed part of the Dufek complex. Whether platinum is also present at depth is another matter altogether – it cannot be presumed to be there solely on the basis of the postulated geological similarity of these two layered igneous bodies.

How reliable is the comparison of the much younger Dufek Massif (180 million years) with the Bushveld (2000 million years)? In the 30 years since it was discovered, reconnaissance geological surveys of the Dufek have mapped the few per cent of the structure which is exposed, aeromagnetic surveys have shown the probable extent beneath the ice, and there have been some rudimentary geochemical studies. From these we can say with certainty that it is a layered, basic, intrusive, igneous complex; the Bushveld is also a layered, basic, intrusive, igneous complex, and has been thoroughly delineated by geological, geophysical and geochemical surveys, drilling and mining, and has been in commercial production for many years. Elsewhere there are other similar, but smaller, layered intrusions – such as the Skaergaard in Greenland, the Sudbury complex in Ontario and the Duluth complex in Minnesota – and in all of these, as with the Bushveld, the economically important minerals occur in the basal part of the structure. The basal part of the Dufek is quite unknown, and at its most accessible point is probably at least 1000 m below the surface (in the deepest part of the sequence it is buried to a depth of at least 6000 m).

There is no *a priori* reason why the Dufek complex should have a basal mineralised zone rich in chromite and platinum and, it must be admitted, no reason why it should not. But presence alone is not enough; we then need to know the thickness and extent of the chromite, and the ore grade and lateral extent of the platinum horizons, all of which can only be established by drilling. One borehole might find the mineral-rich horizon, but many thousands of metres of drilling (certainly not less than thirty thousand) would be required to establish the thickness, extent and richness of the ore body (always remembering that what would constitute an economic deposit in Antarctica will be vastly different from the definition applicable in Australia or Canada). That said, it cannot be denied that if the Dufek intrusion as presently known (and stripped of its ice cover) was located on any other continent, it would rank as a mineral prospect worthy of detailed investigation, and would by now have probably been worked over by generations of geologists and prospectors.

Mining in the Antarctic

How valid is it to keep emphasising the thick covering of ice as a major factor inhibiting exploration and exploitation of minerals should they be present in 'commercial' quantities in Antarctica? After all, there are commercial mining operations in the Canadian Arctic islands (the Polaris Mine on Little Cornwallis Island), and in Greenland (the Black Angel Mine), in which the problems of ice, snow, permafrost, low temperatures and remoteness (accessibility by sea is only possible for six weeks in the year in places) have been successfully overcome. But this is the Arctic, and the Polaris and Black Angel mines are both close to water, whereas the Dufek complex is on the ice sheet some 550 km inland.

Based on this Arctic mining experience, de Wit, in a most imaginative, original and beguilingly confident study,[8] has attempted to extrapolate from the Canadian Arctic to the Antarctic, and puts forward a seemingly plausible design for a platinum mine in the Dufek Massif. He succeeds in demonstrating that if a high value mineral was present near the surface of the ice cover (and this is a very big 'if'), and in large quantities in a deposit with very high assay values, then it could be mined in Antarctica. His case is hypothetical since no such deposit of high value minerals has yet been found, and there are serious doubts that the economic return could ever justify the investment. De Wit may be seriously underestimating the logistic difficulties of, and the environmental constraints on, an Antarctic mining operation, but all the same he has shown how it could be done if expense were no object. There always remains the possibility that if a rich deposit of a high value mineral, of strategic importance and in scarce supply (or to which access on the market was artificially restricted), were to be discovered in Antarctica, normal economic considerations might not obtain and an interested party (perhaps governmental) might elect to underwrite its exploitation come what may.

Hydrocarbons

To date the known occurrences of hydrocarbons are confined to gas encountered in cores obtained by the Deep Sea Drilling Project (DSDP) from the Ross Sea in 1973. This gas was mainly methane, which can be generated in any sedimentary succession which contains organic debris. Why then, on the basis of so little evidence, is it widely believed that if there was to be any exploration for mineral resources in Antarctica, the search would be aimed initially at finding hydrocarbons? Is there perhaps a tacit assumption that mineral exploration is most likely to be a land-based activity which the prevalence of ice would inhibit, whereas oil exploration would be offshore where, seasonally at least, parts of the continental shelf are ice-free?

For the petroleum geologist the one, indeed the only, encouraging factor is the known existence of large sedimentary basins on the continental shelf of Antarctica. The Ross, Weddell, Bellingshausen and Amundsen Seas and the Amery basin are all in this category. However, little is known of these basins other than their areal extent and approximate thickness, and only in the most generalised way can one extrapolate from what is known of the geology onshore. For example, in the geological past the island arc of West Antarctica was the scene of a number of volcanic episodes, and the resulting deposits have been eroded from the Antarctic Peninsula over millions of years and redeposited in deep marine basins on either side of the arc. Geophysical surveys suggest that in places these sediments may be up to 15 km thick, and their geological setting is such that they could possibly contain hydrocarbons. Note the phrase 'geological setting', for once again one is thrown back on a general comparison with the known geology elsewhere in the world, and it must be emphasised that in the Antarctic basins we know nothing of the nature and composition of the sediments, tectonic history or the thermal regime to which the rocks in the basin have been subjected. It is in this context that we must assess all statements about the hydrocarbon reserves of Antarctica.

The requirements for a petroleum province are a thick sedimentary succession of marine origin including source rocks for hydrocarbons, reservoir rocks in which oil and/or gas can accumulate, impermeable rocks to act as a barrier to further migration, and structures that can contain and concentrate the hydrocarbons. In addition, for the genesis and maturation of hydrocarbons from the source rocks, the basin must have been subjected to critical conditions of temperature and pressure in the course of its evolution. Since we only have rudimentary data on the areal extent and sediment thickness in Antarctica, and an absence of all other data, estimates of the hydrocarbon resources of Antarctica are without scientific foundation and at best can only be categorised as guesses.

Such guesses are not uncommon in the literature – following the discovery of the Prudhoe Bay oilfield in Arctic Alaska in 1969, extravagant estimates of the potential of oil and gas reserves of the North American Arctic were published. They too must rank as optimistic guesses, for to date, and notwithstanding the expenditure of hundreds of billions of dollars on exploration, there is still no commercial oil or gas production from the Canadian Arctic islands, or offshore in the Beaufort Sea (and our interpretation of the geology of both these areas is founded on a state of knowledge several orders of magnitude better than that available for Antarctica). Based on the present state of knowledge of Antarctic geology, it is hard to see how Antarctica could come anywhere but last on a global ranking of known sedimentary basins in terms of their petroleum potential.

Regardless of whether or not there are petroleum prospects in Antarctica, it is of very great scientific importance to further the study of its geology, the better to understand the other southern continents and the intervening ocean basins. This can only be achieved by geological surveys of the surface exposures in parallel (and integrated) with geophysical seismic reflection and refraction surveys, and measurements of the earth's gravitational and magnetic fields. The same instruments, methodology and technology are used for scientific research as for commercial exploration, the difference between the two being largely a matter of motive, scale, time, and money. Even though it will be enormously expensive, it is in everybody's interest that the effort devoted to the scientific study of Antarctic geology should be greatly increased, while at the same time maintaining the commitment to the universal availability of raw data, and the widest possible dissemination of results in scientific journals.

This can only be achieved by an enhanced co-operative scientific programme using expertise from many countries, and which is primarily process-orientated and focused on particular problems in those geographic regions which are deemed to offer the best chance of a solution. In pursuit of this objective, co-operation will be the order of the day; although this is generally good and getting better, such large-scale investigative programmes will be complex to fund, manage and support in the field, and in order to succeed, their design and implementation should not be constrained by any limitations of national programmes mounted from national bases.

The long-line geological and geophysical traverse proposed along a line from the Pole northwards approximately along longitude 90° W (see p. 62) is an example of such a project. It would investigate the boundary between Greater Antarctica and Lesser Antarctica, several of the continental blocks and the intervening basin, and would integrate geology and geophysics as, by Antarctic standards, there is good outcrop control from scattered mountain ranges and isolated nunataks along the line of the proposed transect. Where outcrops are poor, drilling is essential in order to establish the nature and age of the stratigraphic succession, and to determine the sonic properties of the rocks, in order to obtain a geological interpretation of the reflection seismic records.

A deep borehole on James Ross Island, for example, would contribute greatly to, and indeed would be a prerequisite for, the geological interpretation of seismic surveys of the Weddell Sea basin. Drilling such a well would be very costly, but even so it would best be undertaken by the Treaty nations rather than by commercial interests, otherwise the scientific motivation for such a project will be open to suspicion by non-Treaty nations (it probably would be anyway, but the risks to the advancement of Antarctic geology would be less). If the monies are not forthcoming for major scientific projects such as long-line transects and deep drilling, Antarctic geology will suffer and the work will never be done until a commercial explorer enters the lists. If and when that ever happens a mineral convention will be in force and, in terms of the advancement of Antarctic science, one of its most important provisions will be that which deals with confidentiality. In framing such a convention the Treaty parties will have to decide which data are the exclusive property of the operator, which data must be furnished to the regulatory authority, and most important of all, how long the regulatory body keeps them confidential before releasing them into the public domain for use by the scientific community at large.

Onshore exploration for hydrocarbons

The petroleum potential of Antarctica is guesswork. All we know with any certainty of the geology as it relates to hydrocarbons is that there are thick offshore sedimentary basins, and that much of the continental shelf dates from the Cretaceous to the Tertiary. For the time being we can ignore the land areas of Antarctica for three good reasons. First, on the basis of the present evidence none of the exposed areas (which total some 140 000 km², or less than half the area of the British Isles) which have been incompletely mapped by reconnaissance geological surveys is worth exploring. Second, seismic surveys through very thick ice are difficult, so it would not be easy to determine the geology beneath the ice, and hence to locate sub-ice structures which would warrant drilling. Lastly, even if structures could be located, wells have never been drilled through a surface section comprising 3–4 km of ice, and moving ice at that. We must therefore conclude that the technology for exploring for and exploiting onshore discoveries in areas with thick permanent ice is not at present available.

It is worth emphasising the significance of drilling in the exploration process; it is important to understand that it is the only way to find oil and gasfields (and indeed many other mineral deposits). The instrument which can detect oil in the ground has yet to be discovered, and the process of exploration is a compound of careful data acquisition involving geological surveys, topographic and bathymetric surveys, geophysical surveys (gravity, aeromagnetic and seismic), and logical appraisal and evaluation of all the available information. Only after this can a prospect be identified, at which point the only way to find out if the structure contains oil or gas is to drill a hole in it. Even if a hole is dry it yields valuable information which may lead to a revised and improved geological interpretation, which is then used to design a more detailed seismic survey, which may reveal new targets worth testing by drilling. This process continues until either hydrocarbons are discovered, or the accumulated evidence is so totally negative that the search is abandoned. World-wide, only one exploration well in fourteen finds oil, and when this sort of success ratio is combined with the enormous estimated cost of an Antarctic well, it is little wonder that oil industry interest in the area is lacking.

Offshore exploration for hydrocarbons

Having ruled out onshore exploration of the ice-covered and, on present evidence, the ice-free parts of the continent, that leaves the offshore basins as the only possible petroleum prospects. These basins were discovered in the course of marine seismic reflection surveys and aeromagnetic studies carried out by a number of countries since 1976. This experience has shown that offshore geological surveys in the Antarctic are a practical proposition with existing technology, and that, given the paucity of geological data, the initial search for potentially drillable structures will be wholly dependent upon marine reflection seismology. Once a structure large enough to be worth drilling has been identified – and in Antarctica 'large enough' would be gigantic anywhere else – the next step would be to design and drill a well.

Do we have the technology to drill an offshore exploration well in Antarctica, either from a floating ice shelf (such as that which covers more than half the Ross Sea), or in areas that are ice-free for a part of the year? In attempting to answer this question one must first consider the risks and problems associated with the physical environment offshore in the Antarctic. These are discussed below with reference to a specific example – the Ross Sea – because, on the present evidence, this area would in all probability be the least difficult in which to begin offshore exploration.

The Ross Sea has all the risks and problems associated with Arctic offshore exploration, plus the additional complication of very deep water; to date drilling in the high Arctic (Beaufort Sea) has been confined to shallow inshore waters, 25–30 m being the maximum water depth tackled to date. In the Ross Sea, water depths are typically in the range 400–600 m, and nowhere are they less than 200 m. Arctic experience has shown the importance of wind and waves, both in terms of the physical loading of structures and as agents of erosion. For example, strong winds blowing across vast areas of sea ice exert an enormous drag stress, which builds up over hundreds of square kilometres and is then concentrated on the drilling structure. Crushing and failure of the ice (or the drilling rig) are the limiting factors in the build-up of these powerful forces; in the Beaufort Sea artificial gravel islands have therefore been used for drilling because a conventional drilling platform would be far too flimsy to withstand these drag factors. The wave climate in the Beaufort Sea is moderate compared to the Ross Sea, which has violent wind and wave conditions as a consequence of the frequent development of depressions round the perimeter of the continent, together with katabatic rushes of cold air outwards from the ice sheet.

Obviously the risk of dynamic impact from icebergs is also considerable; any drilling rig off the Labrador coast of Canada runs the risk of being struck by an iceberg weighing up to 100 million tonnes drifting at up to two knots. These icebergs originate from the Greenland ice sheet and do not drift into the Beaufort Sea, which, however, has its own problems in terms of risk of impact from massive floes of old, highly consolidated multi-year pack ice. In Antarctica, although the ocean is frozen solid for nine months of the year (with pack ice 1.0–2.5 m thick), the sea-ice conditions are marginally better. The same cannot be said of the icebergs; in Antarctica they are commonly 700 m across and 250–400 m thick, and tabular icebergs in excess of 150 km by 70 km have been recorded.[9] As a result of experience in the Arctic, methods have been developed for dealing with the icebergs encountered there, but these techniques cannot be extrapolated to Antarctica to deal with its huge icebergs, and the only option is to make the drilling rig capable of very quick and safe evacuation of the drill site.

A further very serious risk identified in the Arctic is sea-bed scouring or gouging, due to icebergs or heavily keeled multi-year ice floes running aground in water shallower than their draft. Trenches in the sea bed up to 10 m deep can be formed by this process. This has an important bearing on the design of sub-sea pipelines and well-heads, which must be buried below the depth of maximum scour to ensure their integrity. In Antarctica, although grounded icebergs have been seen in the middle of the Ross Sea, intensive scouring appears to be mainly confined to the near-coastal regions.

Given the problems of drilling in offshore waters, would an exploration programme on one of the ice shelves fringing the continent be a more practical proposition? In Antarctica there are extensive ice shelves; practically half of the Ross Sea is covered by the giant Ross Ice Shelf – a 400 m thick, 750 km square expanse of floating ice roughly the size of France. It originates from the polar ice sheet and flows seawards at a rate of 1–5 m per day, accumulating snow on its upper surface and thinning under its own weight. The huge tabular icebergs previously described are calved from the leading edge of the shelf. The water column between the base of the ice and the sea bed only amounts to about ten metres.

On the ice shelf itself the rate of movement constitutes a serious problem for drilling; at five metres per day the drill site would travel almost two kilometres per year relative to the entry point on the sea bed, and there is no way of accommodating this degree of movement within an ice section hundreds of metres thick. Drilling on the moving ice shelf is, however, not completely out of the question. Within the ice shelf there are a few places where the ice has grounded over a shoal to form ice 'rises', which have their own independent (approximately radial) flow regimes, and on the central summit of which the ice is virtually stationary. These ice rises are therefore promising sites for a more or less ice-hazard free year-round drilling operation. Unfortunately, ice rises are not common, and given the enormous area of the ice shelf the odds are against one coinciding with a potentially prospective sub-surface geological structure. Without such a lucky coincidence drilling through the Ross Ice Shelf would hence appear to be out of the question.

The only remaining option, therefore, is drilling from a floating platform in the Ross Sea away from the shelf. The waters of the Ross Sea are navigable every year by ice-strengthened ships during the period from mid-December to mid-March, and a conventional dynamically-positioned drill ship with an ice-strengthened hull could undertake up to three months' exploratory drilling every year; in short, the technology for deep-water drilling in relatively ice-free areas is available on a part-year basis.

Problems of exploitation

So far so good, but there is no point in finding hydrocarbons unless they can be produced economically, with uninterrupted year-round production as the objective. Experience in the Arctic has yielded little or nothing that could be extrapolated to Antarctic operations. There, in the Beaufort Sea, exploration activity has concentrated on near-shallow waters; if a commercial oilfield were to be discovered in this area it would most probably be developed from gravel islands, or some combination of gravel and steel caisson. If this solution were to be adopted in the water depths obtaining in the Ross Sea an absurd quantity of gravel would be required (about five cubic kilometres at a rough estimate!), with no obvious source of the material in the first place. Another possibility would be to install a floating production system kept on station by a fleet of ice-breakers, but the severe winter ice and prevailing wind and wave conditions rule this out. The most probable solution would involve a wholly sub-sea system with well-heads buried to at least ten metres (to protect against iceberg scour), and which would have to function for at least nine months of the year without surface maintenance (power supply/generation would be a problem), or be regularly serviced by submarines. The technology for such a system does not exist at present.

Assuming all the difficulties could be overcome and a safe, reliable production system installed, the problems of storage and export facilities remain to be solved. The point was made earlier that the Arctic consists of frozen ocean surrounded by continents, whereas the Antarctic is a frozen continent surrounded by oceans, and furthermore with no indigenous population. In the Arctic, if oil were to be discovered offshore it would only have to be piped ashore and thence overland to the centres of consumption to the south. In the Antarctic there would seem little point in bringing the oil ashore by pipeline, for there is then nowhere to go but back out to sea again. In any event, pipelining would be difficult because of scouring in coastal waters, and once ashore the oil would have to be stockpiled, which would require the construction of storage and terminal facilities from which it would have to be exported in a burst of activity when the ice breaks up during the brief summer shipping window. Another possibility would be to construct a submarine pipeline north-wards from the production site into waters away from the ice, with year-round access. Such a line would be perhaps 1500 km long in water depths of up to 1000 m, and nothing on this scale has previously been attempted.

Perhaps the most viable method of export from the Ross Sea would be to use ice-breaking tankers (Class VII or better) which could handle year-round ice conditions.[10] If the field were in constant production the ice-breaking traffic would keep a channel substantially open, and the one major technical problem would be to design a free-floating riser which could not be damaged by sea and icebergs, and which ice-breaking tankers could readily find and hook-up to to take on a cargo of oil; yet another item to add to the growing list of theoretical solutions to developing, producing, and exporting petroleum from Antarctica.

So far this account has been almost exclusively concerned with the science and technology of Antarctic mineral resources, but any explor-ation or exploitation requires manpower and materials. The logistic problems of setting up, supplying and supporting a single well are considerable, but they would be as nothing compared to what would be needed to develop an oil or gasfield in the event of a discovery. The difficulties of supporting troops in the field over very long lines of communication were apparent during the Falklands war, and any oilfield development in Antarctica would also involve very long supply lines and encounter many of the same problems (but at least the operation would not have to be carried out under fire). The oil industry has a lot of experience of operating in remote areas with long supply lines, and Alaska in the winter of 1968–69 is a good example of the problems involved. In that case, the oil industry mounted the largest civilian airlift in history, with tens of thousands of tonnes of fuel and equipment for drilling and construction being ferried over a 1000 km supply line from Fairbanks to the shores of the Arctic Ocean as part of a massive exploration drive. The justification for this huge operation was the fact that oil had been discovered and the prospects looked good for the emergence of a new petroleum province; for some the prize was indeed great – the Prudhoe Bay field, largest oilfield on the North American continent – but the majority were disappointed. It is very unlikely, but not totally impossible, that Antarctica would ever attract such an old-fashioned oil-rush.

Prudhoe Bay is an onshore field, as are all the producing fields in the Arctic to date. Since Prudhoe was discovered, the search in the North American Arctic has moved offshore with limited success, and although there have been a number of discoveries in shallow water in the Beaufort Sea, so far none of these have been brought into production. As a result there is no offshore oilfield operating experience which could be applied in Antarctica, and even if there were, the physical conditions of the two areas differ so markedly that the problems of exploring for and producing hydrocarbons in Antarctica will almost inevitably require unique solutions specific to the area.

Environmental considerations

The environmental aspects of operating in Antarctica have so far been mentioned only in passing. The Antarctic is a globally-important, environmentally sensitive area, and it is certain that the standards for safety, pollution prevention and environmental protection will be much higher than the already high standards which apply to Arctic operations. It is right and proper that environmental conservation should be a major consideration, even though it might at times and in places be a constraint upon exploration and development. This concern applies not just to oil exploration or mineral prospecting, but also to the sub-structure for research support, and all of the present activities in Antarctica, such as construction of bases and airstrips, research drilling, etc., are in this category.

Fears are frequently expressed about the possible ecological consequences of an oil spill in Antarctica; once again there is no applicable Arctic experience to go by. Paradoxically this is a penalty of success, because there have been no major oil spills in the Arctic, and hence there are no proven techniques for dealing with major pollution incidents resulting from well blow-outs or pipeline failures under ice. Research has been done on the problem – the Baffin Island Oil Spill project, a co-operative research project of a group of oil companies and the Canadian government, is a notable example – but as the amount of oil permitted to be spilled in field trials of techniques for dealing with oil spills is invariably small, there must be some doubt about extrapolating the results to large pollution incidents. As there have been none of these, the theory and practice of oil spill containment and pick-up under ice has yet to be validated. Clearly prevention is the best form of cure, and high engineering standards and good operating practices have ensured that it has worked so far in the Arctic. No system can be made completely safe and it would be irresponsible and misleading to give the impression that it can be, and while the chance of a major pollution accident can be minimised, it can never be completely eradicated.

Conclusions

This chapter did not set out to down-grade or write off the mineral prospects of Antarctica, but on present evidence there is little or no justification for optimism. This conclusion might be good news for die-hard preservationists, but it will not cause experts in the petroleum or minerals industries to lose any sleep. This is not another justification for the notion if there are any mineral resources in Antarctica they should be forever shut-off from exploitation; future generations may need them and the decision whether or not to explore and develop should be theirs.

Hydrocarbons are a finite, non-renewable resource, and for that reason the large offshore sedimentary basins of Antarctica will one day be explored. That day is very far in the future, and before it arrives other known and more accessible non-conventional sources, such as oil-shales and tar-sands, will be explored. Nor should it be forgotten that only about 40 per cent of the oil in petroleum reservoirs (the recoverable reserves of an oilfield) is presently retrieved; the rest is locked in the pores of the rock in which it is contained, and the search is on for a means of getting it out. The prize for success in the pursuit of enhanced oil recovery is immense because it would vastly increase the recovery from existing fields almost literally overnight. As a corollary it would banish any consideration of Antarctica as a petroleum province even farther into the future. Given the present economic circumstances, with oil in surplus and falling in price, it is hard to envisage any likelihood of oil exploration in Antarctica in the next twenty years – and if the research on enhanced oil recovery is successful, that figure could well stretch to 50 years or more.

In the meantime, whether or not Antarctica ever attracts commercial exploration, it is important at least to maintain, or better still expand, the level of all Antarctic research. In parallel with this research, the principles and practice of environmental protection management should be established and rigorously applied to all Antarctic activities; the experience thus gained would be invaluable in managing resource development if and when oilmen or mineral prospectors ever set foot on the continent. In this and certain other respects the management of Antarctica is too loose for comfort, and now is the time to take action to make a good thing even better. The achievements under the first 25 years of the Treaty are remarkable by any standards. It is greatly to be hoped that the present members and future accessionaires will not just maintain, but further enhance the already high quality of their stewardship.

7

Military potential

From whatever point of view one approaches the question of possible military interest in Antarctica, either from that of the Antarctic Treaty nations or that of the non-parties, it is as well to start by looking at what is forbidden under the Treaty, and to assume that what is not expressly forbidden is permitted.

Article I of the Treaty is all-embracing: 'Antarctica shall be used for peaceful purposes only. There shall be prohibited, *inter alia*, any measure of a military nature, such as the establishment of military bases and fortifications, the carrying out of military manoeuvres, as well as the testing of any type of weapon.' Article V addresses itself specifically to at least one aspect of nuclear weapons: 'Any nuclear *explosions* in Antarctica . . . shall be prohibited' [emphasis added]. Article VII insists that advance notice be given of 'any military personnel or equipment intended to be introduced . . . into Antarctica', in effect a confidence-building measure. However, for the purposes of what follows perhaps the most relevant is Article VI: 'The provisions of the present Treaty shall apply to the area south of 60° South Latitude . . . but nothing in the present Treaty shall prejudice or in any way affect the rights . . . of any State under international law with regard to the high seas within that area.'

One might therefore start by asking whether Antarctica, as defined, is an area in which naval vessels, and in particular nuclear-powered ballistic-missile submarines (SSBNs), are permitted to operate. It might reasonably be assumed that they were, on the basis of Article VI overriding the provisions of Articles I and VII. To date this has not happened (so far as is known), because it has made no military sense for the nuclear powers to use Antarctica as a place in which to conceal SSBNs. This might not always be the case, however, as the following (not unduly alarmist) scenario may serve to illustrate.

Both superpowers are acquiring sea-launched ballistic missiles (SLBMs) of longer and longer range. Within perhaps twenty years they may develop new generation SLBMs with sufficient range (say 8000 nautical miles) to threaten the other's homeland from Antarctic waters. Both superpowers are actively engaged in anti-submarine warfare (ASW) research and development. While no one would yet say that SSBNs have become completely, or even unduly, vulnerable in the open oceans (because submarine detection techniques remain quite modest against SSBNs trying to avoid detection), the possibility of an ASW break-through, perhaps using space-based sensors, clearly exists. If it were to come about that either superpower became very nervous about the survivability of what are, after all, key components of their strategic nuclear forces, either or both might start to think about better con-cealment.

This thought is already leading the Soviet Union, which probably has a higher regard for the ASW capabilities of the United States than vice versa, to contemplate, and even to use, the Arctic ice pack as a place under which to hide SSBNs. There is plenty of evidence of design features incorporated into modern Soviet SSBNs, such as specially strengthened upperworks and retractable fins, which would permit breaking upwards through the thin ice leads in order to fire. The advantages of hiding under ice are obvious; space-based sensors would be blinded, and any attempt at underwater tracking by hunter–killer submarines (SSNs) would be greatly complicated by the existence of ice keels and by the large amount of underwater noise created by pack ice movement, which masks submarine-generated noise. Nevertheless, this is unlikely to prevent the US Navy from attempting to locate and neutralise Soviet SSBNs in time of war, and future American SSNs are likely to be given a greater capacity to operate under polar ice than hitherto.

It is, therefore, by no means unimaginable that superpower interest in the Antarctic ice shelves might grow, despite the fact that it will always take a long time to get there, and it is also not obvious that the Treaty would prevent such SSBN or SSN activity south of 60° S. Perhaps it is worth reflecting that in 1959, when the Treaty was drawn up, the SSBN was in its infancy; the ranges of the early SLBMs were comparatively short, and all the boats afloat or contemplated at the time had to close the coast of the enemy in order to come within firing range. The technological advances since then may simply have been beyond the imaginings of the drafters of the Antarctic Treaty. Alternatively, one might speculate that even then it was superpower insistence that secured the inclusion of the permissive Article VI, against some possible future interest in Antarctic waters.

This is simply one example of an issue that might arise, and which might be worth probing under Article XII. It may also be worth noting that if the superpowers were considering deploying submarines in the area, even in the very distant future, it is certain that they would be interested in Antarctic hydrography, an activity presumably permitted under the Treaty since it is not obviously military in nature.

The provisions for policing the Treaty by the Parties are in place and are generally uncontentious. Article VII of the Treaty provides that each of the Consultative Parties may designate observers, who shall have complete freedom of access at any time to any or all areas of Antarctica. The United States last exercised this power between January and March 1983, and found that 'all nations visited (fourteen stations belonging to a total of eight States) were complying with both the provisions and the spirit of the Antarctic Treaty and its agreed measures'.[1] This was the seventh inspection since 1961. Nothing untoward was found, the only weapons being a flare pistol and a total of nine miscellaneous small arms, which were justified on the grounds of safety or for shooting seals for dog food.

Other views on militarisation

Beyond the issue of SSBN patrol areas it is quite hard – despite a diligent search of the literature – to find any useful references to possible military interest in Antarctica. Indeed, even the possible deployment of SSBNs would appear to be quite fanciful, since virtually no previous commentators seem to have raised it as an issue. There have been the usual smattering of alarmist statements with absolutely no supporting evidence, for example: 'almost certainly, the Russians are interested in Antarctica's . . . strategic significance' (under the title 'Russia's ring around Antarctica').[2] In another article we read:[3] 'this unusual continent may hold yet undetermined military potential', although the article also makes the quite sensible point that it would be better for the United States to remain in Antarctica because 'the military potential of the continent might be realized by others if we [the United States] pulled out of the Treaty and the continent'. Professor Garry McKenzie becomes even more alarmist in this same article when he writes that 'the USSR might take over and use the area to dominate the Southern Ocean' (quite to what purpose, given that there is not much of military interest in the Southern Ocean, he does not explain), and he goes on to cite British interest in preventing German U-boat bases from being established there during the Second World War. He is the one author, albeit in an unsupported statement, to draw attention to Soviet submarine interest: 'The Soviets might find under-ice activities for submarines similar to those in the Arctic Ocean [sic].'

In the same vein, and in the baldest possible manner, Ernst Friedrich Jung states that Antarctica 'is of the greatest strategic value for both world powers, especially with regard to naval warfare and its infrastructure'.[4] Finally, by way of negative evidence, the FCO Background Brief 'Antarctic Treaty, 25th Anniversary' is totally silent on the military question.[5]

Thus, we might reasonably conclude that there is actually not much to say, beyond ritual but unsubstantiated obeisance to the idea that the West ought to prevent the Russians from turning Antarctica into a military bastion for some unnamed purpose.

On the other hand two authors make reference in sensible ways to the time at which the Treaty was signed, and to the climate of the late 1950s. Evan Luard notes: 'There was also a fairly general desire to prevent the area being used for military purposes; serious suggestions had been floated that Antarctica might be a suitable area for nuclear testing or even for the emplacement of missiles.'[6] Deborah Shapley also makes some valid points.[7] She notes that 'military technology has evolved so as to make the Antarctic even less important strategically than in 1961' (surely correct), citing the fact that since then 'the United States has become less dependent on overseas bases' (true) and that 'in the last 23 years satellites have become less dependent on networks of ground tracking-stations, lessening the need to use Antarctica to track satellites' (also true). She believes that 'the [US] Navy has regarded Antarctica as unsuitable for ship or submarine ports' ever since 1946–47, when the US Submarine *Sennett* had to be towed free of Antarctic ice during Operation Highjump. Finally, in an important and sensible conclusion, she notes that 'the Antarctic Treaty has had an advantage over other arms control accords, in that evolving technology has created less rather than more pressure to violate or change it'.

Consequences of technological change

Let us test the above conclusion in a general way against the kind of technological changes that might be anticipated. There are perhaps three areas (besides SSBN patrol patterns) that might be of interest in this context: strategic defences, anti-satellite developments, and ballistic missile trajectory adjustment. To an extent all three are interdependent; all are highly conjectural.

With respect to strategic defences, a preliminary but important consideration to note is that there is very little point in placing a land-based defence or warning system anywhere other than approximately beneath the trajectory of an attacking missile. This rules out Antarctica fairly effectively, except for Fractional Orbital Bombardment Systems (FOBS). There was a time in the 1960s when the United States believed that the Soviet Union was contemplating the use of ballistic missiles to attack the continental United States by 'going round the wrong way'; that is to say instead of using the shortest natural (Arctic) polar orbit, it was thought that the Soviet Union might try to evade American warning systems and ballistic missile defences by attacking from the south. In this eventuality early warning of an attack might have been provided by radars based in Antarctica. However, the Second Common Understanding to Paragraph 2 of Article VII of SALT II requires the USSR to destroy or dismantle twelve of the eighteen FOBS launchers existing in 1979, and to convert the remaining six to 'launchers for testing missiles undergoing modernisation'. So far as is known, this has been done. It has certainly not been raised as a compliance issue by the United States. Thus, so long as FOBS continue to be presumptively banned, Antarctica is uninteresting for land-based defensive purposes.

With regard to the space-based elements of possible strategic defences, it is almost certainly the case that neither superpower will be interested in polar orbits, simply because these are not very useful for intercepting missiles in their most vulnerable boost-phase, and it is upon boost-phase interception that any comprehensive strategic defence would ultimately rely. The only orbits which it makes sense to use for this purpose are inclined orbits which permit maximum time over areas of interest, i.e. missile silos. Thus ground stations would also be out of place in Antarctica. Nor, to make a rather obvious point, would it make any sense to place high-energy lasers on the ground in Antarctica, not merely for security reasons, but also because it would create the longest reflecting path to the Soviet Union via mirrors in space, leading to severe loss of beam intensity at the target. So, for want of any additional evidence to the contrary, it seems reasonable to assert that Antarctica has no relevance to the strategic defences of either superpower. All the activities that might come to matter seem certain to take place in the northern hemisphere.

Anti-satellite activity (the attempts to destroy hostile satellites in orbit) raises some more interesting questions. Although most satellites do not orbit the poles directly, there is one orbit of interest which does. Soviet *Molniya* communications and strategic early warning satellites are placed in highly elliptical orbits with their perigee (lowest point) over Antarctica, so as to spend most of their orbits (eight hours out of twelve) in line-of-sight of the Soviet Union. Although they are travelling at their fastest as they cross the Antarctic, they are at an altitude of only 400–600 km, and this could therefore be their most vulnerable point in orbit to an attacker. Thus it might be tempting, if the American F-15 ASAT system (an aircraft-mounted interceptor missile) is developed and deployed, to move some aircraft to Antarctica in time of war.[8] Apart from this case there is no obvious reason to base an ASAT system in Antarctica rather than any other part of the globe, and many reasons not to do so. This is a very marginal case, but one which permits one to qualify any assertion that there could be no conceivable interest in Antarctica for ASAT purposes. It could be as well to note this possibility if the United States were ever to think in terms of sweeping the skies clear of Soviet satellites.

Finally, it is necessary to consider what was described earlier as 'ballistic missile trajectory adjustment'. If it were to be the case that either superpower were to be faced with significant strategic defences barring the way to transpolar attack in the northern hemisphere, then naturally they might consider again other trajectories. The Soviet Union might once more contemplate FOBS as one way of threatening the continental United States. The underlying assumption here is that the Soviet Union, faced with American strategic defences, might decide not to act co-operatively with the United States in moving from an offence-dominated to a defence-dominated world, but would rather seek to maintain its own offensive capability in the face of American defences. This is likely to require a break-out from SALT constraints. FOBS could then again become an option.

How the United States would react to such a challenge is highly uncertain, and is not something to which much, if any, thought has yet been given. All that can be said in a rather preliminary way is that FOBS would be as vulnerable in its boost phase as any other land-based ballistic missile. Missiles would still lift off from the Soviet Union, however, once free of the atmosphere, the post-boost 'bus' would be travelling in a quite different direction from that covered by the second and third layers of American space-based defences, and it might be harder to engage. When the re-entry vehicles of the FOBS began to descend on the United States after a semi-orbit over the South Pole, they would be coming from a quite different direction than the 'traditional' Soviet re-entry vehicles, thereby greatly complicating American terminal defences which would be looking northwards. Thus the Soviet Union might find that a relatively small number of FOBS could add significantly to the complexity (and so to the cost) of American strategic defences. In the earlier stages of American defensive deployments, the effect of FOBS would be marginal indeed, but the closer the United States came to a truly comprehensive defence against 'traditional' Soviet missiles on 'traditional' trajectories, the more might the Soviet Union become interested again in FOBS.

The above argument contains a large number of assumptions and is too hypothetical to cause anxiety now, but it is necessary to flag it as one of the ways in which pressure might develop on the Antarctic regime. Deborah Shapley does make a connection between FOBS and the Antarctic.[7] She clearly implies that American interest in Antarctica had at one time something to do with FOBS when she states that 'the idea of this Fractional Orbital Bombardment System faded, leaving Antarctica useless for missile tracking'. All that one can do is to reverse the connection and say that if the Soviet Union should again become interested in FOBS, the United States might again see Antarctica as useful for missile tracking.

Some conventional force issues

One obvious reason for wishing to sustain the present regime in Antarctica is the clear possibility that countries desiring to assert competing territorial claims might resort to force. Apart from the superpowers, who have considerable capabilities to back up claims by force using maritime power (and here one must recognise that the United States, mainly with its aircraft carriers, is in a far better position to project military power into Antarctica than is the Soviet Union), it is only Chile and Argentina (and, to a much lesser extent, Australia) which can bring any significant military power to bear in Antarctica. This is as much a fact of geography as anything else. Argentina and Chile are just within range of Grahamland from air bases (Ushuaia and Punta Arenas) in the extreme south of their respective territories, as is the United Kingdom from its new airport at Mount Pleasant in the Falkland Islands. Argentina has one (elderly) aircraft carrier; Chile lacks this ability to carry aircraft closer to Antarctica. Neither would have the capacity for running anything like a continuous war in Antarctica but each could, at a pinch, make life more or less intolerable for the other in isolated outposts. The threat of air attack, in particular, would force a massive – and very costly – investment in air defences, much as Argentina has forced the United Kingdom to invest in air defence in the Falkland Islands. Continuous combat, on the other hand, is virtually unimaginable, and the only conclusion which it seems reasonable to draw is that neither Argentina, Chile nor the United Kingdom could effectively defend their overlapping claims against a determined attempt to prevent occupation. If they draw the same conclusions, all governments must come to realise that only by compromise and accommodation can any of them retain any hold in Antarctica, because each could effectively deny the others the precarious toe-holds needed to maintain claims in a very severe climate.

The resolution of the Beagle Channel dispute between Argentina and Chile, without conflict and on the basis of a reasonable compromise (arbitrated by the Vatican), may be encouraging for the resolution of their conflicting claims to sectors of Antarctica. Indeed, implicit in that settlement might appear to be an acceptance by both countries of the other's legitimate Antarctic claims for the demarcation of the maritime zones, which would seem to have important consequences for those claims. It is not the purpose of this paper to suggest the shape of a compromise between Chile and Argentina (or the United Kingdom) in Antarctica, but it is worth noting here both the cost to all of engaging in a shooting war in Antarctica, and the precedent of the peaceful settlement of the Beagle Channel dispute. Jack Child, Associate Professor of Spanish and Latin American Studies at the American University, has noted that 'the strong Argentine reaction [in 1977] was not so much over the islands themselves as the threat that their possession by Chile posed for the bi-oceanic principle and Argentina's Antarctic claim'.[9]

One final reflection, sometimes raised in Western naval circles, is that Cape Horn, Drake Passage and the Magellan Strait could come to assume very great importance in a protracted East–West war in which, for whatever reason, the Panama Canal was denied to the United States. This region becomes a critical transit area under these, admittedly rather unlikely, circumstances. The possibility of Soviet bases in Grahamland which could threaten American shipping rounding Cape Horn and using Drake Passage could become part of an American nightmare, although it is worth remarking that all current Soviet activity (except for Bellingshausen) is a very long way from Drake Passage. That is another excellent reason for the United States at least to wish to preserve the current demilitarised regime in Antarctica.

The country which more than any other appears to connect the security condition of Antarctica with its own is Australia. The Australian Government has been at pains to point out that the continued demilitarisation of the continent remains a substantial factor in assuring the security of Australia. Australia maintains a lively interest in anything that occurs there that might infringe the Treaty, witness an exchange in the Australian Parliament between Mr Tickner and the Minister of Defence, Mr Beazley, that took place on 28 March 1985 (Question No. 547), relating to satellite receivers in Antarctica. On this occasion Mr Beazley assured his questioner that nothing untoward was taking place and that all satellite monitoring facilities were for scientific purposes (despite claims to the contrary by Greenpeace).

It may be worth adding some reflections for the United Kingdom based on recent experience in South Georgia, a bleak but still much more hospitable environment than Antarctica. It is worth recalling that Britain 'lost' South Georgia in 1982 as comprehensively and as easily as it 'lost' the Falkland Islands. Argentina then lost South Georgia back to Britain with hardly a shot being fired. The point to be made is that, in inhospitable and very isolated regions, it is extraordinarily difficult to hold on to outposts – much more difficult than it is to dispossess the owner thereof. Retention of outposts in the face of a military challenge, and a relatively local military challenge at that, is very difficult and hugely expensive, and could well turn out to be impossible in the long run. Hence Britain would be deluding itself if it believed that it could fight for its Antarctic claims. Thus it would be in the British interest to ensure that issues in Antarctica are resolved peacefully, and that the continent remains demilitarised.

Conclusions

Given that Antarctica is currently demilitarised, any invitation to speculate about its future is bound to move towards (and even beyond) some pretty remote hypotheses. One cannot say that there are no conceivable strategic circumstances in which military interest in Antarctica might not reappear, and for that reason it would seem highly desirable to maintain a regime which seriously inhibits all military deployments in time of peace. Moreover, as Harry Almond has noted:[10] 'the guarantee of demilitarisation is not through enforcement measures, but by the implied threats of each of the parties that if the area is violated, then they will respond in kind'. In that case the existing regime would seem to impose quite adequate constraints, and there is no good reason to meddle with it. Were that regime to be abolished or allowed to lapse, the superpowers at least might flirt with the idea of some marginal militarisation 'just in case'. However, there seems to be no pressing strategic need for either superpower to begin such a military colonisation of Antarctica, and it would be as well to take note of Harry Almond's warning that 'demilitarised arenas are arenas in which states have no reason to compete . . . Once states perceive the need to engage in the arena for their own self-defense, or for establishing a critical base of power, the demilitarised status cannot last'.[11]

At least one reason for asserting that there is no pressing need for militarisation is the fact that even the superpowers would not seek to use Antarctica if the relevant military function could be performed adequately elsewhere. The cost and difficulty of establishing military facilities on the Antarctic continent are likely to be at least one, and possibly even two, orders of magnitude greater than in any more temperate region. Thus the security imperative driving either superpower to attempt a breach of existing arrangements would have to be very great –

and much greater than is reasonably imaginable. Moreover, both the United States and the Soviet Union would know (as noted earlier), that abrogation of the Treaty or an insistence on changing the current rules would let in the other.

Although the question of SSBN safe havens in Antarctica may be remote, it should be recognised that the Treaty does not appear to ban the deployment of SSBNs in the region. It is for others to consider whether the issue should be raised or left fallow. Given the extreme difficulty of policing a rule which kept out SSBNs, and given the general conflict of such a rule with the Law of the Sea and Rights of Passage, it may be judged best not to raise it. We must recognise too that in war the superpowers will act in their own security interests regardless of the provisions of international agreements.

It would be foolish to deny that there could be territorial squabbles over the delimitation of zones which might lead to shooting, especially if the pressures on the current regime come from resource disputes. Nor can one do much to prevent shooting if, say, any of the interested parties came to blows over Grahamland. All that one can usefully say on this point is that all countries will be less likely to be quick on the trigger if there is an international prohibition on the use of force in Antarctica. Such behavioural norms are useful and should be maintained.

It is therefore hard indeed to sustain the alarmist assertions quoted earlier. There would seem to be an excellent chance that the demilitarisation clauses of the Antarctic Treaty could be renewed 'on the nod', and there appears to be no good reason why any State party to the Treaty would wish to argue for revision, and no reason at all why a non-party should do so.

Part III

The future

Since the first decade of its existence the United Nations has devoted much time and effort to the achievement of two objectives, namely decolonisation and universality. Both have now, to all intents and purposes, been achieved. The European empires have disappeared, giving way to an array of newly independent States in Africa, Asia and the Caribbean. The founding membership of the United Nations in 1945 of 51 States has risen to the present total of 159. Where there were originally twelve African and Asian member States, there are now no fewer than 90. With the exception of one or two disputed territories and unresolved problems of self-determination such as the two Koreas and Namibia, it is true to say that every State in the world which feels itself capable, in terms of economic base and size of population, of sustaining the obligations of membership, is now part of the international community.

It can therefore be assumed that, in the decades to come, there will be no significant change in the composition of the United Nations. By the same token it can also be assumed, although with less absolute certainty, that the present politico-economic groupings into which the world is divided will persist, namely the Group of 77 (G77) (which comprises 127 developing countries, including all 101 members of the politically important Non-Aligned Movement (NAM)), Group B (roughly speaking the members of the Organisation for Economic Co-operation and Development (OECD)), and Group D (the Soviet Union and the States of Eastern Europe).

These evolutions have brought about a profound change in the international climate, particularly where economic development and the exploitation of natural resources are concerned. The majority of States, mainly newly independent, embrace some of the poorest countries in the world as well as many in a relatively advanced state of development. They do not take kindly to the perpetuation of the financial and developmental institutions (the IMF being a prime example) which were established by a small group of industrialised countries at a time when they themselves were in no position to participate, and which moreover operate on a system of weighted voting. They favour instead institutions with universal membership in which each State has equal rights. It is significant in this context that even the UN Economic and Social Council (ECOSOC), although designated in the UN Charter as one of the principal organs of the United Nations, along with the General Assembly (GA) and the Security Council, has, because of its limited membership (54 States), been bypassed in the North/South Dialogue which has instead taken place in 'Committees of the Whole' such as the UN Conference on Trade and Development (UNCTAD) and the General Assembly itself.

The States members of the Group of 77 have often emphasised that their objective is the establishment of a 'New International Economic Order'. Among the basic elements of the NIEO, as expressed in the Declaration on 1 May 1974, are affirmations that all States are juridically equal, and that, as equal members of the international community, they have the right to participate fully in the international decision-making process as applied to the solution of world economic, financial, and monetary problems. This implies that decisions should be taken on a one-State-one-vote basis. For many States in the Group the UN Convention on the Law of the Sea (UNCLOS), based in part on a concept of the sea-bed beyond national jurisdiction as the 'common heritage of mankind', and providing for institutions in which decisions are to be taken on a one-State-one-vote basis, was the first multilateral treaty to give legal expression to the NIEO.

This background, the protracted, arduous and contentious negotiations leading to the conclusion of UNCLOS and the failure of the recent attempt to launch Global Negotiations on the whole range of North/South issues have inevitably coloured the view of the Group of 77. It is scarcely surprising then that international attention should have been directed to the Antarctic Treaty, an institution founded by a limited number of States (originally twelve, now eighteen) in which the power of decision-making is confined to those members (the Consultative Parties) who maintain an active presence in Antarctica. In a nutshell, the initiative launched in the United Nations in 1982 by Malaysia has as its objective the replacement of the 'exclusive' Antarctic Treaty System by a universalised regime under the aegis of the United Nations. A detailed case for such a change is argued in Chapter 2 of this study and there is no need to repeat it.

There is equally no need to recapitulate the counter-arguments in favour of the maintenance of the Treaty System as it has evolved over the past 25 years. These are elaborated in Chapter 1. It is also not surprising that many of the Treaty membership should be resistant to change. The scars left by the acrimony of the North/South Dialogue and the exactions of the UNCLOS negotiations are still open. Many of the Treaty members now feel a deep allergy towards multilateral negotiation of yet another universal institution with all that this would involve in terms of confrontation of ideological principle, complicated negotiating procedures, and the prospect of ultimate deadlock or the creation of a new mechanism so labyrinthine as to be inoperable. 'If it works, leave it alone' is the watchword of the Treaty membership.

There is no need either to repeat in detail the views of non-governmental environmentalist groups, which are set out in Chapter 3. Broadly speaking, the more moderate of such groups (as opposed to the more flamboyant bodies which are inclined to over-state and over-glamorise their cases, perhaps in the interests of fund-raising) are favourably disposed towards the Treaty provided that it meets certain challenges, viz. they would like to see improved transmission of information between the Treaty and the outside world, generally freer access to Treaty deliberations, and the maintenance of a vigorous conservation policy.

It is appropriate at this point to analyse the present diplomatic situation in more detail before looking at the possible future course of events. It is important to emphasise at the outset that, although the issue has been debated at the last three sessions of the General Assembly, and although consensus broke down in the voting on the three-part GA Resolution 40/156 of December 1985, Antarctica has not yet developed into a major international confrontation with all that this can involve in terms of angry rhetoric, block voting and the poisoning of inter-State relations on unrelated issues.

In fact, it is clear that there is a great deal of common ground between the critics of the Treaty (the Malaysia Group for short), and its protagonists (the Membership for short). First, everyone wishes to see Antarctica preserved as a genuine 'zone of peace', free from militarisation, nuclearisation or any manifestation of superpower confrontation. These dangers are underlined in Chapter 7, where it is pointed out that, with the advance of military technology since the Treaty was founded in 1959, Antarctic waters could possibly be used by nuclear-powered ballistic missile submarines and, should SALT II collapse without replacement and ballistic missile defence be deployed on a large scale, the southern polar orbit could again become relevant for Fractional Orbital Bombardment Systems, which might, in turn, focus superpower interest on Antarctica.

Second, no one wishes to see the, in some cases, overlapping national claims to segments of Antarctica revived. Again, Chapter 7 adumbrates the dangers of conventional conflict to which the Falklands war between the United Kingdom and Argentina (whose Antarctic claims overlap) has added emphasis.

Third, as argued in Chapters 4 and 5, the importance of continued free and unfettered scientific activity and of the preservation of the unique Antarctic environment unites the international scientific community in favour of a continuing regime in Antarctica which will safeguard these objectives, with the minimum of bureaucratic and diplomatic top hamper.

Fourth, no one wishes to contemplate a commercial free-for-all in terms of mineral (including hydrocarbon) or living resources exploitation (or, for that matter, tourism), which might well put at risk political, environmental and scientific structures.

Hence, the disagreement at this stage appears to be more about means than ends, and the desire to avoid confrontation and the possible collapse of the present system, leaving a vacuum, has been manifest in the moderation of the statements delivered in the United Nations by the Malaysia Group. All speakers have paid tribute to the achievements of the Treaty and have stressed that what they are seeking is not to dismantle it but to extend it into an instrument compatible with the composition and aspirations of the international community of the 1980s and 1990s, as opposed to those of the 1950s and 1960s.

It is also the case that the Malaysia Group, although it secured over 80 votes for GA Resolution 40/156 in 1985, does not command the unanimous support of the NAM/G77 States. A number of important NAM/G77 members, such as India, Cuba, Brazil and Peru have acceded to the Treaty in recent years, while Argentina and Chile are founder-members. This is a further factor which militates against the development of a full-scale diplomatic confrontation. Moreover, the fact that China, although not formally a member of either the Group of 77 or the Non-Aligned Movement, has joined the Treaty adds a major weight to the scales on the side of the Membership.

It would be tempting to conclude from this analysis that the dispute is likely to grumble on in the annual sessions of the General Assembly without either side bringing it to a head. There are many 'hardy annuals' on the Assembly agenda which fall into this category. However, there are three elements in this particular equation each of which could disturb such an uneasy equilibrium; the fact that it will be open in 1991 for any of the Consultative Parties to propose a review of the Treaty, the fact that South Africa is a member (a founder, Consultative member) of the Treaty, and the negotiations under way between the Consultative Parties for a minerals regime (the catalyst which helped to stimulate the Malaysian initiative in the first place).

Options for the future

What of the future? The possibility that the Malaysia Group will lose interest and abandon their campaign can, I believe, be discounted. It has already become part of the fabric of the General Assembly agenda as well as of the proceedings of regional bodies such as the Organisation of African Unity (OAU) and the South Pacific Forum, not to mention the Non-Aligned Movement itself. Furthermore, belief in the principles of universality and the 'common heritage of mankind', fortified by the suspicion that, at some stage in the future, there may be major exploitable quantities of oil and other minerals in Antarctica, is bound to keep the pressure directed at the Membership. Similarly, for all the reasons advanced in earlier chapters and reiterated explicitly or implicitly in this chapter, it is inconceivable that interest in Antarctica on the part of the Membership will wane.

There appear to be four more likely options, each of which needs to be discussed in terms of probability and/or desirability:

Option 1: Both sides stand firm on their present positions in an atmosphere of increasing pressure from the Malaysia Group and united resistance to change on the part of the Membership.

On the face of it, the Membership is in a strong position. It includes the five Permanent Members of the Security Council plus many powerful regional States, for example Japan, Brazil, both Germanies, Italy, Spain and India. Although the total membership in 1986 is only 32 States (18 Consultative Parties and 14 non-Consultative), the Membership represents 80 per cent of the world's population. Against such an array of power there is little effective leverage which the Malaysia Group can exert to force concessions.

However, this option rests on the maintenance of unity amongst the Membership, which is even now less than totally monolithic. Here the South Africa factor could be crucial. It must already be embarrassing in general terms for the NAM/G77 Treaty members to be divided from a majority of their partners, and this discomfort must be considerably heightened by association in the Treaty with the government and representatives of South Africa. This has already manifested itself in that, although the Membership remained virtually united in non-participation in the vote on Part A (further study) and Part B (minerals regime) of GA

Resolution 40/156, they failed to do so over Part C (call for the exclusion of South Africa from Consultative Meetings).

Unless the government in Pretoria has a radical change of heart (highly unlikely), the pressure for further international measures (sanctions) against South Africa is going to increase rapidly. Equally, the situation within South Africa has deteriorated considerably even since the GA Resolution was voted on. It is going to become very difficult for many of the Membership, not only those from the NAM/G77, to continue to sit down at Consultative Meetings with South African representatives. And there is no way, given the consensus rule in the Treaty, by which the South African government can legally be either suspended or excluded without their consent, especially since South Africa maintains a permanent scientific station in Antarctica.

This problem could well undermine the unity of the Membership on which the pursuit of this option depends and, if there is no solution to the South African crisis in the meantime, it could lead to withdrawals and even to the break-up of the Treaty when the time for possible review comes in 1991.

Option 2:   The Membership agree to negotiate a universal regime to cover all aspects of Antarctica on lines analogous to UNCLOS.

This is most unlikely to happen. The pressure from the Malaysia Group for so revolutionary a change is not great, at this stage at any rate, and the degree of resistance amongst the Membership is high. Quite apart from the general fear that such a move would lead to the collapse of the Treaty without an agreed replacement, with all that this would involve, the claimant States would not be prepared to surrender their claims to sovereignty in favour of some form of international trusteeship, particularly in view of the possibility, however remote and far distant, of commercial benefits flowing from Antarctica. The scientific community, already restive at the increasing involvement of lawyers, diplomats, politicians and administrators in the present system, would resist the imposition of what they would see as being a stifling international bureaucracy. It is hard to think of any element in the Treaty community, governmental or non-governmental, which would not strongly oppose such a radical change.

Option 3:   The Malaysia Group swamp the Treaty by mass accessions to it.

The new members would not of course have Consultative status and could not therefore directly influence decision-making. However, such a formidable pressure group on the sidelines of Consultative Meetings could not be ignored and would make it difficult for the present Consultative Parties indefinitely to resist changes.

Nevertheless, this option too seems unlikely to come to pass because of the South African factor. It should be noted that not one of the 50 member States of the OAU has acceded to the Treaty, and that the majority of the Malaysia Group consists of African States and non-African Islamic States. They would not be disposed, simply for tactical advantage, to join an organisation of which South Africa is a member, especially in circumstances where Part C of GA Resolution 40/156 cannot be implemented.

Option 4: That discussions, informal to begin with and without prior commitment, be opened between a limited number of representatives chosen by the Malaysia Group on the one hand, and the Membership on the other, to discover whether an accommodation can be reached on middle ground between Options 1 and 2.

This course of action would appear to be not only the most likely but also the most desirable. The Membership justly pride themselves (see Chapter 1) on the ability of the Treaty System, proven over the past 25 years, to adapt itself to changing circumstances in Antarctica and to the far wider scale of activity and greatly increased permanent and temporary human occupation of the continent. Surely such a flexible organisation should be capable of adapting itself to the concurrent changes which have taken place over the same period in the size and nature of the international community.

The Malaysia Group has already espoused the doctrine that the Treaty should be built upon, rather than dismantled and reconstructed from scratch. These respective attitudes should provide a basis for agreed minimal changes which could revive consensus in the United Nations and avert the danger of a mounting confrontation.

The South African ingredient and its danger to Treaty unity and, in the longer term, to the survival of the Treaty itself, has already been ventilated in the context of the earlier options and it will be referred to again subsequently. Given the crucial present and future importance of the question of potential exploration and exploitation of Antarctic mineral resources, it is appropriate at this stage, before going on to make recommendations, to review the prospects in this regard in the light of Chapter 6 of this study.

Without the advantage of divine inspiration, prophecy is a risky business, particularly where natural resources and the state of the world market are concerned. Who would have predicted, 30 years ago, that the

United Kingdom would shortly be self-sufficient in crude oil with an exportable surplus? And who would have predicted, after the tripling of the price of crude oil in 1973 and its subsequent rise to over $30 per barrel, that, within twelve years, the price would have dropped back to below the December 1973 level? There is a story, familiar to those who have lived and worked in the Middle East, of a petroleum geologist in the 1930s who undertook personally to drink every drop of oil extracted from a certain area: it subsequently turned out to be one of the largest oilfields in the region!

The author of Chapter 6 of this study is well aware of these pitfalls and indeed refers at the outset to a recent, sensational report about offshore oil potential in Antarctica. But he goes on to produce convincing arguments, to this Study Group at least, which suggest that serious exploration for hard-rock minerals (the most important being chromite and platinum) and hydrocarbons, let alone commercial exploitation, is highly unlikely to take place for decades to come, and that the evidence of the existence of worthwhile deposits is still conjectural.

Knowledge of Antarctic geology is in its infancy. Speculation that the Dufek Massif might contain platinum is based purely on a geological comparison with the Bushveld in South Africa (the two are in fact far from analogous). Speculation about the presence of offshore oil and gas derives from the fact that large sedimentary basins are known to exist on the Antarctic continental shelf, as yet unexplored. The author lists the herculean problems involved, which make the exploitation of Arctic oil deposits and the resources of the deep sea-bed seem relatively simple by comparison. Vast distances from potential markets (not a problem in the Arctic), devastating weather conditions which would in the best circumstances limit exploration to three months in each year, the thickness of the ice sheet, the depth of the waters over the continental shelf, massive icebergs, the transportation, storage, and habitation problems particularly in regard to onshore work (the Dufek Massif is 550 km from the nearest open sea) are only some of the obstacles which put Antarctica last in global ranking of interest in the present state of knowledge. There is doubt whether the technology exists even to drill a test well offshore, more so in the case of exploitation. The conclusion is that the oil industry will concentrate on secondary extraction from known oilfields worldwide, and on the exploitation of shale oil deposits and tar sands before they turn their minds to the potential of Antarctica. The possibilities for hard-rock minerals seem equally remote.

Suggestions for the way forward

In all the circumstances analysed in this chapter, I am led to the conclusion that future strains within the Treaty membership, combined with mounting pressure from a significant section of the international community, will make rigid adherence to the *status quo* both undesirable and impracticable, and that both sides would be well advised to consider opening informal negotiations as envisaged in Option 4 above. Such a course of action, if it were to have any hope of success, would necessitate the prior acceptance of certain important propositions.

The Malaysia Group (and the more radical non-governmental organisations which favour notions such as the reservation of Antarctica as a 'World Park') should accept:

1. That a globally administered regime for Antarctica based on the 'common heritage of mankind' is unattainable in present circumstances and should not be demanded. However desirable it may seem in theory to resolve the political and jurisdictional future of Antarctica on universalist lines, the continent differs from, for example, the deep sea-bed, in many important respects, particularly in regard to the existing national claims and counter-claims, the attitude of the powerful non-claimant Treaty members, and the fact that there is already an existing regime in place. In the context of economic development, an inflexible attempt to apply to Antarctica constructs such as the NIEO, or to replicate doctrines inherent in the Law of the Sea Convention would only lead to deadlock, and in the process endanger the existing structure whch has successfully maintained Antarctica as a 'zone of peace' and a unique area of international scientific and environmental co-operation for over a quarter of a century.

2. That the understandable temptation to turn Antarctica into another arena in the battle against apartheid in South Africa should be resisted. In view of the unanimity rule there is no way, except with the (very unlikely) consent of the South African government itself, in which Pretoria can legally be suspended or excluded from the deliberations of the Treaty. Hence all-out pursuit of this objective can only succeed in endangering the Treaty itself, a more important consideration, in the strict Antarctic context, for the world at large than the opening of a fresh sector in the offensive against apartheid.

For their part, the Treaty powers should accept:

1. That all States, not only those which have acceded to the Treaty, have a legitimate interest in the future management of Antarctica, particularly in regard to the insulation of the continent from militarisation and nuclearisation, the commercial exploitation of living and non-living

resources, scientific research in fields which have universal application such as climatology and glaciology, and environmental conservation.

2. That the international politico-economic configuration has changed since 1959 and that it is no longer valid for the Consultative Parties to insist that the rest of the world should leave it to them to manage the functions set out in (1) above and should acquiesce in decisions made without their participation.

3. That, with special regard to the minerals regime, over 50 per cent of the membership of the United Nations are likely to reject any set of regulations which has been drawn up by only eighteen States, however powerful, without the United Nations and/or its relevant Agencies having had an opportunity to participate in their formulation.

On such a basis, it should not be impossible for the two sides to open discussions without prior commitment and with a measure of hope for a positive outcome. The more implacable protagonists of the *status quo* might argue that this course of action would amount to the first step on the slippery slope leading to the morass of another frustrating, inconclusive and, in terms of the future of the Treaty, dangerous multilateral negotiation. The apostles of change might argue that too much is being asked of them in advance by way of the abandonment of important ideological principles. But both would do well to reflect on the likely consequences of perpetuating the present deadlock in an increasingly bitter and menacing international climate.

Assuming that both sides were prepared in principle to contemplate such an exploratory discussion within the framework of the propositions set out above, there are a number of questions which the Consultative Parties would need to consider:

1. *Do the present criteria for Consultative Membership need to be amended, or to be interpreted more flexibly (because of the difficulty of amending the Treaty formally (the unanimity rule, Article XII))?*

Given the origin of the Treaty in the IGY, it was reasonable that the original criterion for Consultative Membership should have been scientific ('. . . demonstrates its interest in Antarctica by conducting substantial scientific research activity there, such as the establishment of a scientific station or the despatch of a scientific expedition' (Article IX)). However, with the development of the Treaty into a comprehensive system for the management of Antarctica, with the conclusion and/or negotiation of regimes designed to govern the commercial exploitation of living and non-living resources, and with the large increase in the numbers of people either permanently stationed in Antarctica or temporarily visiting (tourists and private expeditions), has this narrow criterion now become anachronistic? Should not a State from which, for example, large numbers of tourists visit Antarctica each year, or one whose commercial enterprises are significantly engaged in exploring Antarctic mineral resources, or one whose food supplies involve substantial purchase of krill from Antarctic waters, be as entitled to Consultative Membership as a State whose sole involvement is the establishment of a single scientific station or the despatch of a single annual scientific expedition?

If a less restrictive attitude towards Consultative Membership were to be adopted in order to take into account the widening scale of activity in Antarctica, it should not be impossible to envisage a more internationally acceptable pattern of membership which could be generally compatible with the suggestion in Chapter 2 (pp. 30–3). This would go some of the way towards bridging the gap between the Treaty Membership and their critics.

2. *Is there a need for the creation of a Treaty Secretariat?*

The point has been well made that the Treaty is unique. There certainly can be no other example of an organisation with such a broad range of responsibilities and comprising so large a membership conducting its affairs without any permanent co-ordinating body. There is, of course, no good reason to create an administrative machine where none is required – there are many in the scientific community who believe that the Treaty meetings are already overloaded with diplomats, lawyers and the like – but a small Treaty Secretariat could have two advantages. First, it could provide more co-ordinated follow-up action on the increasing number of decisions made at Consultative Meetings. Second, it could provide a permanent point of contact with the relevant organs of the international community, thus helping to deal with the following problem:

3. *Is there any way in which the present cloud of suspicion, particularly in regard to the negotiation of the minerals regime, can be dispelled?*

Article III of the Treaty provides for the free exchange of scientific observations and results and encourages 'the establishment of co-operative working relations with those Specialized Agencies of the United Nations and other international organizations having a scientific or technical interest in Antarctica'. It is already apparent that the policy of free exchange of information has worked well where purely scientific questions are concerned, but that it comes under strain when potential economic considerations are involved. It is, for example, noted in Chapter 4 that a few countries have refused to release data on the distribution of their fishing fleets and details of fish and krill catches. This is significant. If such national secrecy is maintained between parties to the Treaty itself where commercial advantage is in question, it is scarcely to be wondered at that the negotiation of a minerals regime should have aroused the suspicions of many States and non-governmental environmentalist groups outside the Treaty System. If these suspicions are not to harden into open antagonism and rejection, the Treaty Membership must address what is clearly the nub of the problem, namely to bring about an internationally accepted evolution from being an organisation almost entirely concerned with scientific and environmental matters to one with a significant economic and commercial content. An obvious first step towards defusing suspicion would be to implement the 'establishment of co-operative working relations with . . . Specialized Agencies of the United Nations' by giving such Agencies, as relevant, the right to attend Consultative Meetings (the Membership have already agreed in principle that they should be allowed to attend) and, between meetings, to maintain contact with a Treaty Secretariat if one is established. However, more than this will probably be required and it is appropriate to consider this problem further in the specific context of the future minerals regime. This lies at the heart of the matter:

4. *What considerations should be included in a minerals regime, and with what participation should it be negotiated?*

Assuming that the analysis and conclusions in Chapter 6 are valid, certain further assumptions follow:

1. There is no urgency about the establishment of a minerals regime. There are still some who feel that the current negotiations are an unnecessary source of suspicion and should be deferred or abandoned. However, the general view is that it was prudent for the Membership to start negotiations before they were subjected to commercial and economic pressures. But it is clear that they are in no danger of being overtaken by events.

2. A great deal more needs to be known about the geology of Antarctica, expensive and arduous though the process of ascertainment will be.

3. It is more important, in the light of the length of time available, to get the right minerals regime later, than to get the wrong one sooner. If and when the time comes, public and private corporations are extremely unlikely to show interest unless they can be sure of clear title to a lease or license, with no danger of repercussions against their operations else-

where in the world. This means that an Antarctic minerals regime must be universally accepted with universally endorsed provisions. No regime is likely to work if it is rejected by half the international community (in many of whose States the same potential investors would be active) and kept in place solely by the power of the States comprising the Consultative Membership (India, for example, might be the first of the Consultative Parties to stand against such a policy).

4. The special circumstances of Antarctica – climatic and geographical – will require any enterprise attempting to explore for or to exploit Antarctic mineral resources to be on a massive scale in terms of people and equipment. Because of the importance of conservation of the Antarctic environment, the minerals regime must therefore include far stricter environmental rules than has been the case elsewhere (for example in the Arctic).

5. Very special attention must be given to the questions of jurisdiction and enforcement. Antarctica is a patchwork of national claims, counter-claims, rejected claims and *terra nullius*, all papered over by the Treaty. A resolution of the issues of sovereignty by the establishment of some kind of international trusteeship, however desirable, is impracticable to the point of being illusory. So far, problems of jurisdiction and enforcement have been solved within the framework of the consultative machinery of the Treaty, a task which has become more formidable as the level of permanent habitation and international tourism has increased. The issue is delicate enough where nationals and property of Membership States are concerned; still more so where those of non-Treaty States are in question. Something beyond the present informal *modus vivendi* will be required if and when minerals exploration/exploitation begins. Again, although such arrangements may still have to be informal, in view of the impossibility of resolving the basic issues of sovereignty, they must be universally endorsed if they are to be operationally effective.

Conclusions

There are many ways in which all these considerations could be translated into practice in the framework of discussions within the Membership and between the Membership and the Malaysia Group:

1. The current negotiations for a minerals regime could be given greater transparency by the inclusion in them of a representative of the Secretary-General of the United Nations. This would allay the suspicion aroused by the confidentiality (which in itself seems unnecessary) in which the negotiations have hitherto been conducted and would enhance

the likelihood of the eventual regime receiving international endorsement.

2. A geological survey of Antarctica could be commissioned as an international project, with the co-operation of the Consultative Parties and of the United Nations. The data from such a survey would be made freely available to all governments and to the international scientific and industrial community.

3. The minerals regime negotiated under (1) above should be examined in the light of the results of the survey in (2) above, and no move made to the exploration stage until there was general agreement that all legitimate interests, for example scientific, environmental and developmental, had been taken into account.

4. There should be a further pause for review before the point was reached where exploration gave way to exploitation.

The above suggestions are advanced in an attempt to reconcile the understandable apprehensions of the Membership with the equally understandable aspirations of the Malaysia Group. They may appear to the devotees of the Treaty to introduce perilous complications into the equation, and to its critics that they fall short of their demands. But it should be emphasised first that, since there is no pressing urgency over the minerals question, there is no reason why the negotiating forum should not be made more accessible to the international community and its deliberations more widely disseminated, and also that it is more important to preserve the fundamental elements of the Antarctic Treaty than it is for either side to secure its full objectives. The search for perfection could lead to the collapse of the whole structure. If everyone demonstrates restraint and flexibility there is no reason why Antarctica should follow the rest of the globe down the road of diplomatic confrontation or East/West or North/South antagonism. All parties would, at this important stage of the debate, do well to bear in mind the message of the Treaty that 'it is in the interest of all mankind that Antarctica shall continue forever to be used exclusively for peaceful purposes and shall not become the scene or object of international discord'.

Postscript

In September 1986, the Non-Aligned Summit at Harare included a passage on Antarctica in the Final Declaration. Discussion of this item was limited because of the preoccupation of the Summit as a whole with other major problems including the deteriorating situation in South Africa and because the subject had been so thoroughly discussed at the

Luanda Non-Aligned Foreign Ministers meeting in 1985. The text of the passage reflected the substance of the three General Assembly resolutions adopted in 1985 although the language was vestigially less strong in certain respects. Some observers even thought that it might be possible to restore consensus at the forthcoming General Assembly.

In November, the Chairman of the Ninth Session of the Special Consultative Meeting on Antarctic Mineral Resources held in Tokyo issued a Press Release setting out the state of play on the negotiations for a minerals regime. The statement went into some detail on progress made, and emphasised the fact that the Treaty countries were negotiating 'an open regime' which could be joined by any State that 'accepted the Antarctic Treaty'. The Chairman stressed the part played in the meeting by the non-Consultative parties to the Treaty ('The fact is that any State can now play an effective and constructive role in discussions on Antarctic matters simply by acceding to the Antarctic Treaty.').

The General Assembly debate took place in November. In the event consensus was not restored but the temperature of the debate was no higher than in 1985. The wording of the three resolutions was in all cases slightly stronger than that of the texts at Appendix II. For example, on the minerals regime negotiations, the General Assembly resolution in 1985 invited the Consultative Parties to inform the Secretary-General of the United Nations of negotiations to establish a minerals regime. In 1986 the equivalent resolution called for a moratorium on negotiations until all members of the international community could participate fully in them.

As regards voting, support for the resolution on the provision of information fell slightly, from 96 to 94 votes, while support for that on the minerals regime rose slightly, from 92 to 96 votes. Support for the third resolution – the exclusion of South Africa – rose from 100 to 119 votes, including the positive votes of six of the Consultative Parties (Argentina, Brazil, China, India, Poland, the Soviet Union) as compared with two in 1985.

As regards those Consultative Parties supporting the resolution on South Africa, in 1986 a statement was made in the First Committee, on behalf of the Parties to the Antarctic Treaty, that they would vote on the resolution 'in ways which do not affect their position on the successful functioning of the Antarctic Treaty'.

I do not believe that these subsequent events in any way modify the analyses and judgments already made in Part III. A study of the latest voting figures suggests that, provided the Treaty Membership stands firm and continues to lobby strongly, the Malaysia Group is unlikely to be able to add significantly to the approximately 60 per cent of the total UN membership which now supports the resolution on the minerals regime and the provision of information.* However, as anticipated, support for the resolution on the South African aspect has grown and now includes a substantial proportion of the Consultative Parties. It is on this point that future pressure is likely to develop further.

---

* For example, on the minerals regime 96 States voted in favour. As the total number of votes possible was 159, 63 for one reason or another did not favour the resolution. Of those 63 States, 32 are Treaty Parties and the Treaty 'lobbying', effective for a variety of reasons, makes it unlikely that any of them will change their stance on the issue.

It is possible to make an assessment of the remaining 31 as follows:

7 (Byelorussia, Democratic Yemen, Lao PDR, Mongolia, Nicaragua, Ukraine, Vietnam) are likely to continue to demonstrate sympathy for the USSR position, by anything from non-participation to simple absence;

7 (Colombia, Costa Rica, Ecuador, El Salvador, Guatemala, Paraguay, Venezuela) are likely to follow the lead of the major Latin American powers, Argentina/Brazil, in the same way;

4 (Fiji, Solomon Is., Vanuatu, Samoa) are likely to reflect Australia's stance in some way, while trying not to oppose the majority of the Group of 77;

4 (Greece, Luxemburg, Portugal, Ireland) are likely to reflect, in varying degrees, solidarity with the EEC;

9 Israel's pro-Treaty and Iceland's pro-Scandinavian non-participation stance is not likely to change (2), nor are the sympathies of Austria, Canada and Turkey (3); Bahamas, Dominica, Gambia and Seychelles (4) have been absent during the vote for two consecutive years.

From the foregoing, it would seem that the lines of the conflict have now been drawn, and there is very little room for substantial increases on either side. Perhaps some of the small island States in the last category may change their practice of being absent, and decide to join the majority, but even then the latter cannot reach the (current) two-thirds mark of 106, even if that had a particular relevance.

APPENDIX I

Text of the Antarctic Treaty

The Governments of Argentina, Australia, Belgium, Chile, the French Republic, Japan, New Zealand, Norway, the Union of South Africa, the Union of Soviet Socialist Republics, the United Kingdom of Great Britain and Northern Ireland, and the United States of America,

Recognizing that it is in the interest of all mankind that Antarctica shall continue forever to be used exclusively for peaceful purposes and shall not become the scene or object of international discord;

Acknowledging the substantial contributions to scientific knowledge resulting from international co-operation in scientific investigation in Antarctica;

Convinced that the establishment of a firm foundation for the continuation and development of such co-operation on the basis of freedom of scientific investigation in Antarctica as applied during the International Geophysical Year accords with the interests of science and the progress of all mankind;

Convinced also that a treaty ensuring the use of Antarctica for peaceful purposes only and the continuance of international harmony in Antarctica will further the purposes and principles embodied in the Charter of the United Nations;

Have agreed as follows:

Article I

1. Antarctica shall be used for peaceful purposes only. There shall be prohibited, *inter alia*, any measure of a military nature, such as the establishment of military bases and fortifications, the carrying out of military manoeuvres, as well as the testing of any type of weapon.
2. The present Treaty shall not prevent the use of military personnel or equipment for scientific research or for any other peaceful purpose.

Article II

Freedom of scientific investigation in Antarctica and co-operation toward that end, as applied during the International Geophysical Year, shall continue, subject to the provisions of the present Treaty.

Article III

1. In order to promote international co-operation in scientific investigation in Antarctica, as provided for in Article II of the present Treaty, the Contracting Parties agree that, to the greatest extent feasible and practicable:
(a) information regarding plans for scientific programs in Antarctica shall be exchanged to permit maximum economy of and efficiency of operations;
(b) scientific personnel shall be exchanged in Antarctica between expeditions and stations;
(c) scientific observations and results from Antarctica shall be exchanged and made freely available.
2. In implementing this Article, every encouragement shall be given to the establishment of co-operative working relations with those Specialized Agencies of the United Nations and other international organizations having a scientific or technical interest in Antarctica.

Article IV

1. Nothing contained in the present Treaty shall be interpreted as:
(a) a renunciation by any Contracting Party of previously asserted rights of or claims to territorial sovereignty in Antarctica;
(b) a renunciation or diminution by any Contracting Party of any basis of claim to territorial sovereignty in Antarctica which it may have whether as a result of its activities or those of its nationals in Antarctica, or otherwise;
(c) prejudicing the position of the Contracting Party as regards its recognition or non-recognition of any other State's rights of or claim or basis of claim to territorial sovereignty in Antarctica.
2. No acts or activities taking place while the present Treaty is in force shall constitute a basis for asserting, supporting or denying a claim to territorial sovereignty in Antarctica or create any rights of sovereignty in Antarctica. No new claim, or enlargement of an existing claim, to territorial sovereignty in Antarctica shall be asserted while the present Treaty is in force.

Article V

1. Any nuclear explosions in Antarctica and the disposal there of radioactive waste material shall be prohibited.
2. In the event of the conclusion of international agreements concerning the use of nuclear energy, including nuclear explosions and the disposal of radioactive waste material, to which all of the Contracting Parties whose representatives are entitled to participate in the meetings provided for under Article IX are parties, the rules established under such agreements shall apply in Antarctica.

Article VI

The provisions of the present Treaty shall apply to the area south of 60° South Latitude, including all ice shelves, but nothing in the present Treaty shall prejudice or in any way affect the rights, or the exercise of the rights, of any State under international law with regard to the high seas within that area.

Article VII

1. In order to promote the objectives and ensure the observance of the provisions of the present Treaty, each Contracting Party whose representatives are entitled to participate in the meetings referred to in Article IX of the Treaty shall have the right to designate observers to carry out any inspection provided for by the present Article. Observers shall be nationals of the Contracting Parties which designate them. The names of observers shall be communicated to every other Contracting Party having the right to designate observers, and like notice shall be given of the termination of their appointment.

2. Each observer designated in accordance with the provisions of paragraph 1 of this Article shall have complete freedom of access at any time to any or all areas of Antarctica.

3. All areas of Antarctica, including all stations, installations and equipment within those areas, and all ships and aircraft at points of discharging or embarking cargoes or personnel in Antarctica, shall be open at all times to inspection by any observers designated in accordance with paragraph 1 of this Article.

4. Aerial observation may be carried out at any time over any or all areas of Antarctica by any of the Contracting Parties having the right to designate observers.

5. Each Contracting Party shall, at the time when the present Treaty enters into force for it, inform the other Contracting Parties, and thereafter shall give them notice in advance, of

   (a) all expeditions to and within Antarctica, on the part of its ships or nationals, and all expeditions to Antarctica organized in or proceeding from its territory;

   (b) all stations in Antarctica occupied by its nationals; and

   (c) any military personnel or equipment intended to be introduced by it into Antarctica subject to the conditions prescribed in paragraph 2 of Article I of the present Treaty.

Article VIII

1. In order to facilitate the exercise of their functions under the present Treaty, and without prejudice to the respective positions of the Contracting Parties relating to jurisdiction over all other persons in Antarctica, observers designated under paragraph 1 of Article VII and scientific personnel exchanged under sub-paragraph 1(b) of Article III of the Treaty, and members of the staffs accompanying any such persons, shall be subject only to the jurisdiction of the Contracting Party of which they are nationals in respect of all acts or omissions occurring while they are in Antarctica for the purpose of exercising their functions.

2. Without prejudice to the provisions of paragraph 1 of this Article, and pending the adoption of measures in pursuance of sub-paragraph 1(e) of Article IX, the Contracting Parties concerned in any case of dispute with regard to the exercise of jurisdiction in Antarctica shall immediately consult together with a view to reaching a mutually acceptable solution.

Article IX

1. Representatives of the Contracting Parties named in the preamble to the present Treaty shall meet at the City of Canberra within two months after the date of

entry into force of the Treaty, and thereafter at suitable intervals and places, for the purpose of exchanging information, consulting together on matters of common interest pertaining to Antarctica, and formulating and considering, and recommending to their Governments, measures in furtherance of the principles and objectives of the Treaty, including measures regarding:

(a) use of Antarctica for peaceful purposes only;

(b) facilitation of scientific research in Antarctica;

(c) facilitation of international scientific co-operation in Antarctica;

(d) facilitation of the exercise of the rights of inspection provided for in Article VII of the Treaty

(e) questions relating to the exercise of jurisdiction in Antarctica;

(f) preservation and conservation of living resources in Antarctica.

2. Each Contracting Party which has become a party to the present Treaty by accession under Article XIII shall be entitled to appoint representatives to participate in the meetings referred to in paragraph 1 of the present Article, during such times as that Contracting Party demonstrates its interest in Antarctica by conducting substantial scientific research activity there, such as the establishment of a scientific station or the despatch of a scientific expedition.

3. Reports from the observers referred to in Article VII of the present Treaty shall be transmitted to the representatives of the Contracting Parties participating in the meetings referred to in paragraph 1 of the present Article.

4. The measures referred to in paragraph 1 of this Article shall become effective when approved by all the Contracting Parties whose representatives were entitled to participate in the meetings held to consider those measures.

5. Any or all of the rights established in the present Treaty may be exercised as from the date of entry into force of the Treaty whether or not any measures facilitating the exercise of such rights have been proposed, considered or approved as provided in this Article.

Article X

Each of the Contracting Parties undertakes to exert appropriate efforts, consistent with the Charter of the United Nations, to the end that no one engages in any activity in Antarctica contrary to the principles or purposes of the present Treaty.

Article XI

1. If any dispute arises between two or more of the Contracting Parties concerning the interpretation or application of the present Treaty, those Contracting Parties shall consult among themselves with a view to having the dispute resolved by negotiation, inquiry, mediation, conciliation, arbitration, judicial settlement or other peaceful means of their own choice.

2. Any dispute of this character not so resolved shall, with the consent, in each case, of all parties to the dispute, be referred to the International Court of Justice for settlement; but failure to reach agreement on reference to the International Court shall not absolve parties to the dispute from the responsibility of continuing to seek to resolve it by any of the various peaceful means referred to in paragraph 1 of this Article.

Article XII

1. (a) The present Treaty may be modified or amended at any time by unanimous agreement of the Contracting Parties whose representatives are entitled to participate in the meetings provided for under Article IX. Any such modification or amendment shall enter into force when the depositary Government has received notice from all such Contracting Parties that they have ratified it.

   (b) Such modification or amendment shall thereafter enter into force as to any other Contracting Party when notice of ratification by it has been received by the depositary Government. Any such Contracting Party from which no notice of ratification is received within a period of two years from the date of entry into force of the modification or amendment in accordance with the provision of sub-paragraph 1(a) of this Article shall be deemed to have withdrawn from the present Treaty on the date of the expiration of such period.

2. (a) If after the expiration of thirty years from the date of entry into force of the present Treaty, any of the Contracting Parties whose representatives are entitled to participate in the meetings provided for under Article IX so requests by a communication addressed to the depositary Government, a Conference of all the Contracting Parties shall be held as soon as practicable to review the operation of the Treaty.

   (b) Any modification or amendment to the present Treaty which is approved at such a Conference by a majority of the Contracting Parties there represented, including a majority of those whose representatives are entitled to participate in the meetings provided for under Article IX, shall be communicated by the depositary Government to all Contracting Parties immediately after the termination of the Conference and shall enter into force in accordance with the provisions of paragraph 1 of the present Article.

   (c) If any such modification or amendment has not entered into force in accordance with the provisions of sub-paragraph 1(a) of this Article within a period of two years after the date of its communication to all the Contracting Parties, any Contracting Party may at any time after the expiration of that period give notice to the depositary Government of its withdrawal from the present Treaty; and such withdrawal shall take effect two years after the receipt of the notice by the depositary Government.

Article XIII

1. The present Treaty shall be subject to ratification by the signatory States. It shall be open for accession by any State which is a Member of the United Nations, or by any other State which may be invited to accede to the Treaty with tne consent of all the Contracting Parties whose representatives are entitled to participate in the meetings provided for under Article IX of the Treaty.

2. Ratification of or accession to the present Treaty shall be effected by each State in accordance with its constitutional processes.

3. Instruments of ratification and instruments of accession shall be deposited with the Government of the United States of America, hereby designated as the depositary Government.

4. The depositary Government shall inform all signatory and acceding States of the date of each deposit of an instrument of ratification or accession, and the date of entry into force of the Treaty and of any modification or amendment thereto.

5. Upon the deposit of instruments of ratification by all the signatory States, the present Treaty shall enter into force for those States and for States which have deposited instruments of accession. Thereafter the Treaty shall enter into force for any acceding State upon the deposit of its instruments of accession.
6. The present Treaty shall be registered by the depositary Government pursuant to Article 102 of the Charter of the United Nations.

Article XIV

The present Treaty, done in the English, French, Russian and Spanish languages, each version being equally authentic, shall be deposited in the archives of the Government of the United States of America, which shall transmit duly certified copies thereof to the Governments of the signatory and acceding States.

APPENDIX II

Texts of relevant conventions and UN resolutions

Convention on the Conservation of Antarctic Marine Living Resources

The Contracting Parties,

Recognising the importance of safeguarding the environment and protecting the integrity of the ecosystem of the seas surrounding Antarctica;

Noting the concentration of marine living resources found in Antarctic waters and the increased interest in the possibilities offered by the utilization of these resources as a source of protein;

Conscious of the urgency of ensuring the conservation of Antarctic marine living resources;

Considering that it is essential to increase knowledge of the Antarctic marine ecosystem and its components so as to be able to base decisions on harvesting on sound scientific information;

Believing that the conservation of Antarctic marine living resources calls for international co-operation with due regard for the provisions of the Antarctic Treaty, and with the active involvement of all States engaged in research or harvesting activities in Antarctic waters;

Recognising the prime responsibilities of the Antarctic Treaty Consultative Parties for the protection and preservation of the Antarctic environment and, in particular, their responsibilities under Article IX, paragraph 1(f) of the Antarctic Treaty in respect of the preservation and conservation of living resources in Antarctica;

Recalling the action already taken by the Antarctic Treaty Consultative Parties including in particular the Agreed Measures for the Conservation of Antarctic Fauna and Flora, as well as the provisions of the Convention for the Conservation of Antarctic Seals;

Bearing in mind the concern regarding the conservation of Antarctic marine living resources expressed by the Consultative Parties at the Ninth Consultative Meeting of the Antarctic Treaty and the importance of the provisions of Recommendation IX-2 which led to the establishment of the present Convention;

Believing that it is in the interest of all mankind to preserve the waters surrounding the Antarctic continent for peaceful purposes only and to prevent their becoming the scene or object of international discord;

Recognising, in the light of the foregoing, that it is desirable to establish suitable machinery for recommending, promoting, deciding upon and co-ordinating the measures and scientific studies needed to ensure the conservation of Antarctic marine living organisms;

Have agreed as follows:

Article I

1. This Convention applies to the Antarctic marine living resources of the area south of 60° South latitude and to the Antarctic marine living resources of the area between that latitude and the Antarctic Convergence which form part of the Antarctic marine ecosystem.
2. Antarctic marine living resources means the populations of fin fish, molluscs, crustaceans and all other species of living organisms, including birds, found south of the Antarctic Convergence.
3. The Antarctic marine ecosystem means the complex of relationships of Antarctic marine living resources with each other and with their physical environment.
4. The Antarctic Convergence shall be deemed to be a line joining the following points along parallels of latitude and meridians of longitude: 50° S, 0°; 50° S, 30° E; 45° S, 30° E; 45° S, 80° E; 55° S, 80° E; 55° S, 150° E; 60° S, 150° E; 60° S, 50° W; 50° S, 50° W; 50° S, 0°.

Article II

1. The objective of this Convention is the conservation of Antarctic marine living resources.
2. For the purposes of this Convention, the term 'conservation' includes rational use.
3. Any harvesting and associated activities in the area to which this Convention applies shall be conducted in accordance with the provisions of this Convention and with the following principles of conservation:

   (a) prevention of decrease in the size of any harvested population to levels below those which ensure its stable recruitment. For this purpose its size should not be allowed to fall below a level close to that which ensures the greatest net annual increment;

   (b) maintenance of the ecological relationships between harvested, dependent and related populations of Antarctic marine living resources and the restoration of depleted populations to the levels defined in sub-paragraph (a) above; and

   (c) prevention of changes or minimization of the risk of changes in the marine ecosystem which are not potentially reversible over two or three decades, taking into account the state of available knowledge of the direct and indirect impact of harvesting, the effect of the introduction of alien species, the effects of associated activities on the marine ecosystem and of the effects of environmental changes, with the aim of making possible the sustained conservation of Antarctic marine living resources.

Article III

The Contracting Parties, whether or not they are Parties to the Antarctic Treaty, agree that they will not engage in any activities in the Antarctic Treaty area contrary to the principles and purposes of that Treaty and that, in their relations with each other, they are bound by the obligations contained in Articles I and V of the Antarctic Treaty.

Article IV

1. With respect to the Antarctic Treaty area, all Contracting Parties, whether or not they are Parties to the Antarctic Treaty, are bound by Articles IV and VI of the Antarctic treaty in their relations with each other.
2. Nothing in this Convention and no acts or activities taking place while the present Convention is in force shall:

(a) constitute a basis for asserting, supporting or denying a claim to territorial sovereignty in the Antarctic Treaty area or create any rights of sovereignty in the Antarctic Treaty area;

(b) be interpreted as a renunciation or diminution by any Contracting Party of, or as prejudicing, any right or claim or basis of claim to exercise coastal state jurisdiction under international law within the area to which this Convention applies;

(c) be interpreted as prejudicing the position of any Contracting Party as regards its recognition or non-recognition of any such right, claim or basis of claim;

(d) affect the provision of Article IV, paragraph 2, of the Antarctic Treaty that no new claim, or enlargement of an existing claim, to territorial sovereignty in Antarctica shall be asserted while the Antarctic Treaty is in force.

Article V

1. The Contracting Parties which are not Parties to the Antarctic Treaty acknowledge the special obligations and responsibilities of the Antarctic Treaty Consultative Parties for the protection and preservation of the environment of the Antarctic Treaty area.
2. The Contracting Parties which are not Parties to the Antarctic Treaty agree that, in their activities in the Antarctic Treaty area, they will observe as and when appropriate the Agreed Measures for the Conservation of Antarctic Fauna and Flora and such other measures as have been recommended by the Antarctic treaty Consultative Parties in fulfilment of their responsibility for the protection of the Antarctic environment from all forms of harmful human interference.
3. For the purpose of this Convention, 'Antarctic Treaty Consultative Parties' means the Contracting Parties to the Antarctic Treaty whose Representatives participate in meetings under Article IX of the Antarctic Treaty.

Article VI

Nothing in this Convention shall derogate from the rights and obligations of Contracting Parties under the International Convention for the regulation of Whaling and the Convention for the Conservation of Antarctic Seals.

Article VII

1. The Contracting Parties hereby establish and agree to maintain the Commission for the Conservation of Antarctic Marine Living Resources (hereinafter referred to as 'the Commission').
2. Membership in the Commission shall be as follows:
(a) each Contracting Party which participated in the meeting at which this Convention was adopted shall be a member of the Commission;
(b) each State Party which has acceded to this Convention pursuant to Article XXIX shall be entitled to be a Member of the Commission during such time as that acceding party is engaged in research or harvesting activities in relation to the marine living resources to which this Convention applies;
(c) each regional economic integration organization which has acceded to this Convention pursuant to Article XXIX shall be entitled to be a Member of the Commission during such time as its States members are so entitled;
(d) a Contracting Party seeking to participate in the work of the Commission pursuant to sub-paragraphs (b) and (c) above shall notify the Depositary of the basis upon which it seeks to become a Member of the Commission and of its willingness to accept conservation measures in force. The Depositary shall communicate to each Member of the Commission such notification and accompanying information. Within two months of receipt of such communication from the Depositary, any Member of the Commission may request that a special meeting of the Commission be held to consider the matter. Upon receipt of such request, the Depositary shall call such a meeting. If there is no request for a meeting, the Contracting Party submitting the notification shall be deemed to have satisfied the requirements for Commission Membership.
3. Each Member of the Commission shall be represented by one representative who may be accompanied by alternate representatives and advisers.

Article VIII

The Commission shall have legal personality and shall enjoy in the territory of each of the States Parties such legal capacity as may be necessary to perform its function and achieve the purposes of this Convention. The privileges and immunities to be enjoyed by the Commission and its staff in the territory of a State Party shall be determined by agreement between the Commission and the State Party concerned.

Article IX

1. The function of the Commission shall be to give effect to the objective and principles set out in Article II of this Convention. To this end, it shall:
(a) facilitate research into and comprehensive studies of Antarctic marine living resources and of the Antarctic marine ecosystem;
(b) compile data on the status of and changes in population of Antarctic marine living resources and on factors affecting the distribution, abundance and productivity of harvested species and dependent or related species or populations;
(c) ensure the acquisition of catch and effort statistics on harvested populations;
(d) analyse, disseminate and publish the information referred to in sub-paragraphs (b) and (c) above and the reports of the Scientific Committee;

(e) identify conservation needs and analyse the effectiveness of conservation measures;

(f) formulate, adopt and revise conservation measures on the basis of the best scientific evidence available, subject to the provisions of paragraph 5 of this Article;

(g) implement the system of observation and inspection established under Article XXIV of this Convention;

(h) carry out such other activities as are necessary to fulfil the objectives of this Convention.

2. The conservation measures referred to in paragraph 1 (f) above include the following:

(a) the designation of the quantity of any species which may be harvested in the area to which this Convention applies;

(b) the designation of regions and sub-regions based on the distribution of populations of Antarctic marine living resources;

(c) the designation of the quantity which may be harvested from the populations of regions and sub-regions;

(d) the designation of protected species;

(e) the designation of the size, age and, as appropriate, sex of species which may be harvested;

(f) the designation of open and closed seasons for harvesting;

(g) the designation of the opening and closing of areas, regions or sub-regions for purposes of scientific study or conservation, including special areas for protection and scientific study;

(h) regulation of the effort employed and methods of harvesting, including fishing gear, with a view, *inter alia*, to avoiding undue concentration of harvesting in any region or sub-region;

(i) the taking of such other conservation measures as the Commission considers necessary for the fulfilment of the objective of this Convention, including measures concerning the effects of harvesting and associated activities on components of the marine ecosystem other than the harvested populations.

3. The Commission shall publish and maintain a record of all conservation measures in force.

4. In exercising its functions under paragraph 1 above, the Commission shall take full account of the recommendations and advice of the Scientific Committee.

5. The Commission shall take full account of any relevant measures or regulations established or recommended by the Consultative Meetings pursuant to Article IX of the Antarctic Treaty or by existing fisheries commissions responsible for species which may enter the area to which this Convention applies, in order that there shall be no inconsistency between the rights and obligations of a Contracting Party under such regulations or measures and conservation measures which may be adopted by the Commission.

6. Conservation measures adopted by the Commission in accordance with this Convention shall be implemented by Members of the Commission in the following manner.

(a) the Commission shall notify conservation measures to all Members of the Commission;

(b) conservation measures shall become binding upon all Members of the Commission 180 days after such notification, except as provided in sub-paragraphs (c) and (d) below;

(c) if a Member of the Commission, within ninety days following the notification specified in sub-paragraph (a), notifies the Commission that it is unable to accept the conservation measure, in whole or in part, the measure shall not, to the extent stated, be binding upon that Member of the Commission;

(d) in the event that any Member of the Commission invokes the procedure set forth in sub-paragraph (c) above, the Commission shall meet at the request of any Member of the Commission to review the conservation measure. At the time of such meeting and within thirty days following the meeting, any Member of the Commission shall have the right to declare that it is no longer able to accept the conservation measure, in which case the Member shall no longer be bound by such measure.

Article X

1. The Commission shall draw the attention of any State which is not a Party to this Convention to any activity undertaken by its nationals or vessels which, in the opinion of the Commission, affects the implementation of the objectives of this Convention.
2. The Commission shall draw the attention of all Contracting Parties to any activity which, in the opinion of the Commission, affects the implementation by a Contracting Party of the objective of this Convention or the compliance by that Contracting Party with its obligations under this Convention.

Article XI

The Commission shall seek to cooperate with Contracting Parties which may exercise jurisdiction in marine areas adjacent to the area to which this Convention applies in respect of the conservation of any stock or stocks of associated species which occur both within those areas and the area to which this Convention applies, with a view to harmonizing the conservation measures adopted in respect of such stocks.

Article XII

1. Decisions of the Commission on matters of substance shall be taken by consensus. The question of whether a matter is one of substance shall be treated as a matter of substance.
2. Decisions on matters other than those referred to in paragraph 1 above shall be taken by a simple majority of the Members of the Commission present and voting.
3. In Commission consideration of any item requiring a decision, it shall be made clear whether a regional economic integration organization will participate in the taking of the decision and, if so, whether any of its member States will also participate. The number of Contracting Parties so participating shall not exceed the number of member States of the regional economic integration organization which are Members of the Commission.
4. In the taking of decisions pursuant to this Article, a regional economic integration organization shall have only one vote.

Article XIII

1. The headquarters of the Commission shall be established at Hobart, Tasmania, Australia.
2. The Commission shall hold a regular annual meeting. Other meetings shall also be held at the request of one-third of its members and as otherwise provided in this Convention. The first meeting of the Commission shall be held within three months of the entry into force of this Convention, provided that among the Contracting Parties there are at least two States conducting harvesting activities within the area to which this Convention applies. The first meeting shall, in any event, be held within one year of the entry into force of this Convention. The Depositary shall consult with the signatory States regarding the first Commission meeting, taking into account that a broad representation of such States is necessary for the effective operation of the Commission.
3. The Depositary shall convene the first meeting of the Commission at the headquarters of the Commission. Thereafter, meetings of the Commission shall be held at its headquarters, unless it decides otherwise.
4. The Commission shall elect from among its members a Chairman and Vice-Chairman, each of whom shall serve for a term of two years and shall be eligible for re-election for one additional term. The first Chairman shall, however, be elected for an initial term of three years. The Chairman and Vice-Chairman shall not be representatives of the same Contracting Party.
5. The Commission shall adopt and amend as necessary the rules of procedure for the conduct of its meetings, except with respect to the matters dealt with in Article XII of this Convention.
6. The Commission may establish such subsidiary bodies as are necessary for the performance of its functions.

Article XIV

1. The Contracting Parties hereby establish the Scientific Committee for the Conservation of Antarctic Marine Living Resources (hereinafter referred to as 'the Scientific Committee') which shall be a consultative body to the Commission. The Scientific Committee shall normally meet at the headquarters of the Commission unless the Scientific Committee decides otherwise.
2. Each member of the Commission shall be a member of the Scientific Committee and shall appoint a representative with suitable scientific qualifications who may be accompanied by other experts and advisers.
3. The Scientific Committee may seek the advice of other scientists and experts as may be required on an *ad hoc* basis.

Article XV

1. The Scientific Committee shall provide a forum for consultation and co-operation concerning the collection, study and exchange of information with respect to the marine living resources to which this Convention applies. It shall encourage and promote co-operation in the field of scientific research in order to extend knowledge of the marine living resources of the Antarctic marine ecosystem.
2. The Scientific Committee shall conduct such activities as the Commission may direct in pursuance of the objective of this Convention and shall:

(a) establish criteria and methods to be used for determinations concerning the conservation measures referred to in Article IX of this Convention;

(b) regularly assess the status and trends of the populations of Antarctic marine living resources;

(c) analyse data concerning the direct and indirect effects of harvesting on the populations of Antarctic marine living resources;

(d) assess the effects of proposed changes in the methods or levels of harvesting and proposed conservation measures;

(e) transmit assessments, analyses, reports and recommendations to the Commission as requested or on its own initiative regarding measures and research to implement the objective of this Convention;

(f) formulate proposals for the conduct of international and national programs of research into Antarctic marine living resources.

3. In carrying out its functions, the Scientific Committee shall have regard to the work of other relevant technical and scientific organizations and to the scientific activities conducted within the framework of the Antarctic Treaty.

Article XVI

1. The first meeting of the Scientific Committee shall be held within three months of the first meeting of the Commission. The Scientific Committee shall meet thereafter as often as may be necessary to fulfil its functions.

2. The Scientific Committee shall adopt and amend as necessary its rules of procedure. The rules and any amendments thereto shall be approved by the Commission. The rules shall include procedures for the presentation of minority reports.

3. The Scientific Committee may establish, with the approval of the Commission, such subsidiary bodies as are necessary for the performance of its functions.

Article XVII

1. The Commission shall appoint an Executive Secretary to serve the Commission and Scientific Committee according to such procedures and on such terms and conditions as the Commission may determine. His term of office shall be for four years and he shall be eligible for re-appointment.

2. The Commission shall authorize such staff establishment for the Secretariat as may be necessary and the Executive Secretary shall appoint, direct and supervise such staff according to such rules and procedures and on such terms and conditions as the Commission may determine.

3. The Executive Secretary and Secretariat shall perform the functions entrusted to them by the Commission.

Article XVIII

The official languages of the Commission and of the Scientific Committee shall be English, French, Russian and Spanish.

Article XIX

1. At each annual meeting, the Commission shall adopt by consensus its budget and the budget of the Scientific Committee.

2. A draft budget for the Commission and the Scientific Committee and any subsidiary bodies shall be prepared by the Executive Secretary and submitted to

the Members of the Commission at least sixty days before the annual meeting of the Commission.

3. Each Member of the Commission shall contribute to the budget. Until the expiration of five years after the entry into force of this Convention, the contribution of each Member of the Commission shall be equal. Thereafter the contribution shall be determined in accordance with two criteria: the amount harvested and an equal sharing among all Members of the Commission. The Commission shall determine by consensus the proportion in which these two criteria shall apply.

4. The financial activities of the Commission and Scientific Committee shall be conducted in accordance with financial regulations adopted by the Commission and shall be subject to an annual audit by external auditors selected by the Commission.

5. Each Member of the Commission shall meet its own expenses arising from attendance at meetings of the Commission and of the Scientific Committee.

6. A Member of the Commission that fails to pay its contributions for two consecutive years shall not, during the period of its default, have the right to participate in the taking of decisions in the Commission.

Article XX

1. The Members of the Commission shall, to the greatest extent possible, provide annually to the Commission and to the Scientific Committee such statistical, biological and other data and information as the Commission and Scientific Committee may require in the exercise of their functions.

2. The Members of the Commission shall provide, in the manner and at such intervals as may be prescribed, information about their harvesting activities, including fishing areas and vessels, so as to enable reliable catch and effort statistics to be compiled.

3. The Members of the Commission shall provide to the Commission at such intervals as may be prescribed information on steps taken to implement the conservation measures adopted by the Commission.

4. The Members of the Commission agree that in any of their harvesting activities, advantage shall be taken of opportunities to collect data needed to assess the impact of harvesting.

Article XXI

1. Each Contracting Party shall take appropriate measures within its competence to ensure compliance with the provisions of this Convention and with conservation measures adopted by the Commission to which the Party is bound in accordance with Article IX of this Convention.

2. Each Contracting Party shall transmit to the Commission information on measures taken pursuant to paragraph 1 above, including the imposition of sanctions for any violation.

Article XXII

1. Each Contracting Party undertakes to exert appropriate efforts, consistent with the Charter of the United Nations, to the end that no one engages in any activity contrary to the objective of this Convention.

2. Each Contracting Party shall notify the Commission of any such activity which comes to its attention.

Article XXIII

1. The Commission and the Scientific Committee shall co-operate with the Antarctic Treaty Consultative Parties on matters falling within the competence of the latter.
2. The Commission and the Scientific Committee shall co-operate, as appropriate, with the Food and Agriculture Organisation of the United Nations and with other Specialized Agencies.
3. The Commission and the Scientific Committee shall seek to develop co-operative working relationships, as appropriate, with inter-governmental and non-governmental organizations which could contribute to their work including the Scientific Committee on Antarctic Research, the Scientific Committee on Ocean Research and the International Whaling Commission.
4. The Commission may enter into agreements with the organizations referred to in this Article and with other organizations as may be appropriate. The Commission and the Scientific Committee may invite such organizations to send observers to their meetings and to meetings of their subsidiary bodies.

Article XXIV

1. In order to promote the objective and ensure observance of the provisions of this Convention, the Contracting Parties agree that a system of observation and inspection shall be established.
2. The system of observation and inspection shall be elaborated by the Commission on the basis of the following principles:
    (a) Contracting Parties shall co-operate with each other to ensure the effective implementation of the system of observation and inspection, taking account of the existing international practice. This system shall include, *inter alia*, procedures for boarding and inspection by observers and inspectors designated by the Members of the Commission and procedures for flag state prosecution and sanctions on the basis of evidence resulting from such boarding and inspections. A report of such prosecutions and sanctions imposed shall be included in the information referred to in Article XXI of this Convention;
    (b) in order to verify compliance with measures adopted under this Convention, observation and inspection shall be carried out on board vessels engaged in scientific research or harvesting of marine living resources in the area to which this Convention applies, through observers and inspectors designated by the Members of the Commission and operating under terms and conditions to be established by the Commission;
    (c) designated observers and inspectors shall remain subject to the jurisdiction of the Contracting Party of which they are nationals. They shall report to the Member of the Commission by which they have been designated which in turn shall report to the Commission.
3. Pending the establishment of the system of observation and inspection, the Members of the Commission shall seek to establish interim arrangements to designate observers and inspectors and such designated observers and inspectors shall be entitled to carry out inspections in accordance with the principles set out in paragraph 2 above.

Article XXV

1. If any dispute arises between two or more of the Contracting Parties concerning the interpretation or application of this Convention, those Contracting Parties shall consult among themselves with a view to having the dispute resolved by negotiation, inquiry, mediation, conciliation, arbitration, judicial settlement or other peaceful means of their own choice.
2. Any dispute of this character not so resolved shall, with the consent in each case of all Parties to the dispute, be referred for settlement to the International Court of Justice or to arbitration; but failure to reach agreement on reference to the International Court or to arbitration shall not absolve Parties to the dispute from the responsibility of continuing to seek to resolve it by any of the various peaceful means referred to in paragraph 1 above.
3. In cases where the dispute is referred to arbitration, the arbitral tribunal shall be constituted as provided in the Annex to this Convention.

Article XXVI

1. This Convention shall be open for signature at Canberra from 1 August to 31 December 1980 by the States participating in the Conference on the Conservation of Antarctic Marine Living Resources held at Canberra from 7 to 20 May 1980.
2. The States which so sign will be the original signatory States of the Convention.

Article XXVII

1. This Convention is subject to ratification, acceptance or approval by signatory States.
2. Instruments of ratification, acceptance or approval shall be deposited with the Government of Australia, hereby designated as the Depositary.

Article XXVIII

1. This Convention shall enter into force on the thirtieth day following the date of deposit of the eighth instrument of ratification, acceptance or approval by States referred to in paragraph 1 of Article XXVI of this Convention.
2. With respect to each State or regional economic integration organization which subsequent to the date of entry into force of this Convention deposits an instrument of ratification, acceptance, approval or accession, the Convention shall enter into force on the thirtieth day following such deposit.

Article XXIX

1. This Convention shall be open for accession by any State interested in research or harvesting activities in relation to the marine living resources to which this Convention applies.
2. This Convention shall be open for accession by regional economic integration organizations constituted by sovereign States which include among their members one or more States Members of the Commission and to which the States Members of the organization have transferred, in whole or in part, competences with regard to the matters covered by this Convention. The accession of such regional economic integration organizations shall be the subject of consultations among Members of the Commission.

Article XXX

1. This Convention may be amended at any time.
2. If one-third of the Members of the Commission request a meeting to discuss a proposed amendment the Depositary shall call such a meeting.
3. An amendment shall enter into force when the Depositary has received instruments of ratification, acceptance or approval thereof from all the Members of the Commission.
4. Such amendment shall thereafter enter into force as to any other Contracting Party when notice of ratification, acceptance or approval by it has been received by the Depositary. Any such Contracting Party from which no such notice has been received within a period of one year from the date of entry into force of the amendment in accordance with paragraph 3 above shall be deemed to have withdrawn from this Convention.

Article XXXI

1. Any Contracting Party may withdraw from this Convention on 30 June of any year, by giving written notice not later than 1 January of the same year to the Depositary, which, upon receipt of such a notice, shall communicate it forthwith to the other Contracting Parties.
2. Any other Contracting Party may, within sixty days of the receipt of a copy of such a notice from the Depositary, give written notice of withdrawal to the Depositary in which case the Convention shall cease to be in force on 30 June of the same year with respect to the Contracting Party giving such notice.
3. Withdrawal from this Convention by any Member of the Commission shall not affect its financial obligations under this Convention.

Article XXXII

The Depositary shall notify all Contracting Parties of the following:
(a) signatures of this Convention and the deposit of instruments of ratification, acceptance, approval or accession;
(b) the date of entry into force of this Convention and of any amendment thereto.

Article XXXIII

1. This Convention, of which the English, French, Russian and Spanish texts are equally authentic, shall be deposited with the Government of Australia which shall transmit duly certified copies thereof to all signatory and acceding Parties.
2. This Convention shall be registered by the Depositary pursuant to Article 102 of the Charter of the United Nations.
Drawn up at Canberra this twentieth day of May 1980.
In witness whereof the undersigned, being duly authorized, have signed this Convention.

Signatures and Ratifications

State	Date of signature	Date of deposit ratification, (AP) or acceptance (AC)
Argentina		28 Apr. 1982
Australia		6 May 1981
Belgium	11 Sept 1980	
Chile		22 July 1981
France	16 Sept. 1980	16 Sept. 1982
German Democratic Republic	11 Sept. 1980	30 Mar. 1982
Germany, Federal Republic of		23 Apr. 1982 *
Japan	12 Sept. 1980	26 May 1981 (AC)
New Zealand		8 Mar. 1982
Norway		
Poland		
South Africa	11 Sept. 1980	23 July 1981
Union of Soviet Socialist Republics		26 May 1981 (AP)
United Kingdom		31 Aug. 1981
United States of America		18 Feb. 1982

* Includes Berlin (West).

Accessions

State	Date
European Economic Community	21 Apr. 1982

Annex for an arbitral tribunal

1. The arbitral tribunal referred to in paragraph 3 of Article XXV shall be composed of three arbitrators who shall be appointed as follows:

(a) The Party commencing proceedings shall communicate the name of an arbitrator to the other Party which, in turn, within a period of forty days following such notification, shall communicate the name of the second arbitrator. The Parties shall, within a period of sixty days following the appointment of the second arbitrator, appoint the third arbitrator, who shall not be a national of either Party and shall not be of the same nationality as either of the first two arbitrators. The third arbitrator shall preside over the tribunal.

(b) If the second arbitrator has not been appointed within the prescribed period, or if the Parties have not reached agreement within the prescribed period on the appointment of the third arbitrator, that arbitrator shall be appointed, at the request of either Party, by the Secretary-General of the Permanent Court of Arbitration, from among persons of international standing not having the nationality of a State which is a Party to this Convention.

2. The arbitral tribunal shall decide where its headquarters will be located and shall adopt its own rules of procedure.

3. The award of the arbitral tribunal shall be made by a majority of its members, who may not abstain from voting.

4. Any Contracting Party which is not a Party to the dispute may intervene in the proceedings with the consent of the arbitral tribunal.
5. The award of the arbitral tribunal shall be final and binding on all Parties to the dispute and on any Party which intervenes in the proceedings and shall be complied with without delay. The arbitral tribunal shall interpret the award at the request of one of the Parties to the dispute or of any intervening Party.
6. Unless the arbitral tribunal determines otherwise because of the particular circumstances of the case, the expenses of the tribunal, including the remuneration of its members, shall be borne by the Parties to the dispute in equal shares.

Convention for the Conservation of Antarctic Seals

The Contracting Parties,

Recalling the Agreed Measures for the Conservation of Antarctic Fauna and Flora, adopted under the Antarctic Treaty signed at Washington on 1 December 1969;

Recognizing the general concern about the vulnerability of Antarctic seals to commercial exploitation and the consequent need for effective conservation measures;

Recognizing that the stocks of Antarctic seals are an important living source in the marine environment which requires an international agreement for its effective conservation;

Recognizing that this resource should not be depleted by over-exploitation, and hence that any harvesting should be regulated so as not to exceed the levels of the optimum sustainable yield;

Recognizing that in order to improve scientific knowledge and so place exploitation on a rational basis, every effort should be made both to encourage biological and other research on Antarctic seal populations and to gain information from such research and from the statistics of future sealing operations, so that further suitable regulations may be formulated;

Noting that the Scientific Committee on Antarctic Research of the International Council of Scientific Unions (SCAR) is willing to carry out the tasks requested of it in this Convention;

Desiring to promote and achieve the objectives of protection, scientific study and rational use of Antarctic seals, and to maintain a satisfactory balance within the ecological system,

Have agreed as follows:

Article 1
Scope
1. This Convention applies to the seas south of 60° South Latitude, in respect of which the Contracting Parties affirm the provisions of Article IV of the Antarctic Treaty.
2. This Convention may be applicable to any or all of the following species:
   Southern elephant seal *Mirounga leonina*,
   Leopard seal *Hydrurga leptonyx*,
   Weddell seal *Leptonychotes weddelli*,
   Crabeater seal *Lobodon carcinophagus*,

Ross seal *Ommatophoca rossi*,
Southern fur seals *Arctocephalus* sp.

3. The Annex to this Convention forms an integral part thereof.

Article 2
Implementation

1. The Contracting Parties agree that the species of seals enumerated in Article 1 shall not be killed or captured within the Convention area by their nationals or vessels under their respective flags except in accordance with the provisions of this Convention.

2. Each Contracting Party shall adopt for its nationals and for vessels under its flag such laws, regulations and other measures, including a permit system as appropriate, as may be necessary to implement this Convention.

Article 3
Annexed measures

1. This Convention includes an Annex specifying measures which the Contracting Parties hereby adopt. Contracting Parties may from time to time in the future adopt other measures with respect to the conservation, scientific study and rational and humane use of seal resources, prescribing *inter alia:*
   (a) permissible catch;
   (b) protected and unprotected species;
   (c) open and closed seasons;
   (d) open and closed areas, including the designation of reserves;
   (e) the designation of special areas where there shall be no disturbance of seals;
   (f) limits relating to sex, size, or age for each species;
   (g) restrictions relating to time of day and duration, limitations of effort and methods of sealing;
   (h) types and specifications of gear and apparatus and appliances which may be used:
   (i) catch returns and other statistical and biological records;
   (j) procedures for facilitating the review and assessment of scientific information:
   (k) other regulatory measures including an effective system of inspection.

2. The measures adopted under paragraph (1) of this Article shall be based upon the best scientific and technical evidence available.

3. The Annex may from time to time be amended in accordance with the procedures provided for in Article 9.

Article 4
Special permits

1. Notwithstanding the provisions of this Convention, any Contracting Party may issue permits to kill or capture seals in limited quantities and in conformity with the objectives and principles of this Convention for the following purposes:
   (a) to provide indispensable food for men or dogs;
   (b) to provide for scientific research; or
   (c) to provide specimens for museums, educational or cultural institutions.

2. Each Contracting Party shall, as soon as possible, inform the other Contracting

Parties and SCAR of the purpose and content of all permits issued under paragraph (1) of this Article and subsequently of the numbers of seals killed or captured under these permits.

Article 5
Exchange of information and scientific advice
1. Each Contracting Party shall provide to the other Contracting Parties and to SCAR the information specified in the Annex within the period indicated therein.
2. Each Contracting Party shall also provide to the other Contracting Parties and to SCAR before 31 October each year information on any steps it has taken in accordance with Article 2 of this Convention during the preceding period 1 July to 30 June.
3. Contracting Parties which have no information to report under the two preceding paragraphs shall indicate this formally before 31 October each year.
4. SCAR is invited:
(a) to assess information received pursuant to this Article; encourage exchange of scientific data and information among the Contracting Parties; recommend programmes for scientific research; recommend statistical and biological data to be collected by sealing expeditions within the Convention area; and suggest amendments to the Annex; and
(b) to report on the basis of the statistical, biological and other evidence available when the harvest of any species of seal in the Convention area is having a significantly harmful effect on the total stocks of such species or on the ecological system in any particular locality.
5. SCAR is invited to notify the Depositary which shall report to the Contracting Parties when SCAR estimates in any sealing season that the permissible catch limits for any species are likely to be exceeded and, in that case, to provide an estimate of the date upon which the permissible catch limits will be reached. Each Contracting Party shall then take appropriate measures to prevent its nationals and vessels under its flag from killing or capturing seals of that species after the estimated date until the Contracting Parties decide otherwise.
6. SCAR may if necessary seek the technical assistance of the Food and Agriculture Organization of the United Nations in making its assessments.
7. Notwithstanding the provisions of paragraph (1) of Article 1 the Contracting Parties shall, in accordance with their internal law, report to each other and to SCAR, for consideration, statistics relating to the Antarctic seals listed in paragraph (2) of Article 1 which have been killed or captured by their nationals and vessels under their respective flags in the area of floating sea ice north of 60° South Latitude.

Article 6
Consultations between Contracting Parties
1. At any time after commercial sealing has begun a Contracting Party may propose through the Depositary that a meeting of Contracting Parties be convened with a view to:
(a) establishing by a two-thirds majority of the Contracting Parties, including the concurring votes of all States signatory to this Convention present at the meeting, an effective system of control, including inspection, over the implementation of the provisions of this Convention;

(b) establishing a commission to perform such functions under this Convention as the Contracting Parties may deem necessary; or

(c) considering other proposals, including:

  (i) the provision of independent scientific advice;

  (ii) the establishment, by a two-thirds majority, of a scientific advisory committee which may be assigned some or all of the functions requested of SCAR under this Convention, if commercial sealing reaches significant proportions;

  (iii) the carrying out of scientific programmes with the participation of the Contracting Parties; and

  (iv) the provision of further regulatory measures, including moratoria.

2. If one-third of the Contracting Parties indicate agreement the Depositary shall convene such a meeting, as soon as possible.

3. A meeting shall be held at the request of any Contracting Party, if SCAR reports that the harvest of any species of Antarctic seal in the area to which this Convention applies is having a significantly harmful effect on the total stocks or the ecological system in any particular locality.

Article 7
Review of operations

The Contracting Parties shall meet within five years after the entry into force of this Convention and at least every five years thereafter to review the operation of the Convention.

Article 8
Amendments to the Convention

1. This Convention may be amended at any time. The text of any amendment proposed by a Contracting Party shall be submitted to the Depositary, which shall transmit it to all the Contracting Parties.

2. If one-third of the Contracting Parties request a meeting to discuss the proposed amendment the Depositary shall call such a meeting.

3. An amendment shall enter into force when the Depositary has received instruments of ratification or acceptance thereof from all the Contracting Parties.

Article 9
Amendments to the Annex

1. Any Contracting Party may propose amendments to the Annex to this Convention. The text of any such proposed amendment shall be submitted to the Depositary which shall transmit it to all Contracting Parties.

2. Each such proposed amendment shall become effective for all Contracting Parties six months after the date appearing on the notification from the Depositary to the Contracting Parties, if within 120 days of the notification date, no objection has been received and two-thirds of the Contracting Parties have notified the Depositary in writing of their approval.

3. If an objection is received from any Contracting Party within 120 days of the notification date, the matter shall be considered by the Contracting Parties at their next meeting. If unanimity on the matter is not reached at the meeting, the Contracting Parties shall notify the Depositary within 120 days from the date of closure of the meeting of their approval or rejection of the original amendment

or of any new amendment proposed by the meeting. If, by the end of this period, two-thirds of the Contracting Parties have approved such amendment, it shall become effective six months from the date of the closure of the meeting for those Contracting Parties which have by then notified their approval.

4. Any Contracting Party which has objected to a proposed amendment may at any time withdraw that objection, and the proposed amendment shall become effective with respect to such Party immediately if the amendment is already in effect, or at such time as it becomes effective, under the terms of this Article.

5. The Depositary shall notify each Contracting Party immediately upon receipt of each approval or objection, of each withdrawal of objection, and of the entry into force of any amendment.

6. Any State which becomes a party to this Convention after an amendment to the Annex has entered into force shall be bound by the Annex as so amended. Any State which becomes a Party to this Convention during the period when a proposed amendment is pending may approve or object to such an amendment within the time limits applicable to other Contracting Parties.

Article 10
Signature
This Convention shall be open for signature at London from 1 June to 31 December 1972 by States participating in the Conference on the Conservation of Antarctic Seals held at London from 3 to 11 February 1972.

Article 11
Ratification
This Convention is subject to ratification or acceptance. Instruments of ratification or acceptance shall be deposited with the Government of the United Kingdom of Great Britain and Northern Ireland, hereby designated as the Depositary.

Article 12
Accession
This Convention shall be open for accession by any State which may be invited to accede to this Convention with the consent of all the Contracting Parties.

Article 13
Entry into force
1. This Convention shall enter into force on the thirtieth day following the date of deposit of the seventh instrument of ratification or acceptance.

2. Thereafter this Convention shall enter into force for each ratifying, accepting or acceding State on the thirtieth day after deposit by such State of its instrument of ratification, acceptance or accession.

Article 14
Withdrawal
Any Contracting Party may withdraw from this Convention on 30 June of any year by giving notice on or before 1 January of the same year to the Depositary, which upon receipt of such a notice shall at once communicate it to the other

Contracting Parties. Any other Contracting Party may, in like manner, within one month of the receipt of a copy of such a notice from the Depositary, give notice of withdrawal, so that the Convention shall cease to be in force on 30 June of the same year with respect to the Contracting Party giving such notice.

Article 15
Notifications by the Depositary

The Depositary shall notify all signatory and acceding States of the following:

(a) signatures of this Convention, the deposit of instruments of ratification, acceptance or accession and notices of withdrawal;

(b) the date of entry into force of this Convention and of any amendments to it or its Annex.

Article 16
Certified copies and registration

1. This Convention, done in the English, French, Russian and Spanish languages, each version being equally authentic, shall be deposited in the archives of the Government of the United Kingdom of Great Britain and Northern Ireland, which shall transmit duly certified copies thereof to all signatory and acceding States.
2. This Convention shall be registered by the Depositary pursuant to Article 102 of the Charter of the United Nations.

Annex

1. Permissible catch

The Contracting Parties shall in any one year, which shall run from 1 July to 30 June inclusive, restrict the total number of seals of each species killed or captured to the numbers specified below. These numbers are subject to review in the light of scientific assessments:

(a) in the case of Crabeater seals *Lobodon carcinophagus*, 175,000;

(b) in the case of Leopard seals *Hydrurga leptonyx*, 12,000;

(c) in the case of Weddell seals *Leptonychotes weddelli*, 5,000.

2. Protected species

(a) it is forbidden to kill or capture Ross seals *Ommatophoca rossi*, Southern elephant seals *Mirounga leonina*, or fur seals of the genus *Arctocephalus*.

(b) In order to protect the adult breeding stock during the period when it is most concentrated and vulnerable, it is forbidden to kill or capture any Weddell seal *Leptonychotes weddelli* one year old or older between 1 September and 31 January inclusive.

3. Closed season and sealing season

The period between 1 March and 31 August inclusive is a Closed Season, during which the killing or capturing of seals is forbidden. The period 1 September to the last day in February constitutes a Sealing Season.

4. Sealing zones

Each of the sealing zones listed in this paragraph shall be closed in numerical sequence to all sealing operations for the seal species listed in paragraph 1 of this Annex for the period 1 September to the last day of February inclusive. Such closures shall begin with the same zone as is closed under paragraph 2 of Annex B to Annex 1 of the Report of the Fifth Antarctic Treaty Consultative Meeting at the moment the Convention enters into force. Upon the expiration of each closed period, the affected zone shall reopen:

Zone 1–between 60° and 120° West Longitude

Zone 2–between 0° and 60° West Longitude, together with that part of the Weddel Sea lying westward of 60° West Longitude

Zone 3–between 0° and 70° East Longitude

Zone 4–between 70° and 130° East Longitude

Zone 5–between 130° East Longitude and 170° West Longitude

Zone 6–between 120° and 170° West Longitude.

5. *Seal reserves*

It is forbidden to kill or capture seals in the following reserves, which are seal breeding areas or the site of long-term scientific research:

(a) The area around the South Orkney Islands between 60° 20' and 60° 56' South Latitude and 44° 05' and 46° 25' West Longitude.

(b) The area of the southwestern Ross Sea south of 76° South Latitude and west of 170° East Longitude.

(c) The area of Edisto Inlet south and west of a line drawn between Cape Hallet at 72° 19' South Latitude, 170° 18' East Longitude, and Helm Point, at 72° 11' South Latitude, 170° 00' East Longitude.

6. *Exchange of information*

(a) Contracting Parties shall provide before 31 October each year to other Contracting Parties and to SCAR a summary of statistical information on all seals killed or captured by their nationals and vessels under their respective flags in the Convention area, in respect of the preceding period 1 July to 30 June. This information shall include by zones and months:

 (i) The gross and nett tonnage, brake horse-power, number of crew, and number of days' operation of vessels under the flag of the Contracting Party;

 (ii) The number of adult individuals and pups of each species taken.

When specially requested, this information shall be provided in respect of each ship, together with its daily position at noon each operating day and the catch on that day.

(b) When an industry has started, reports of the number of seals of each species killed or captured in each zone shall be made to SCAR in the form and at the intervals (not shorter than one week) requested by that body.

(c) Contracting Parties shall provide to SCAR biological information, in particular:

 (i) Sex

 (ii) Reproductive condition

 (iii) Age

SCAR may request additional information or material with the approval of the Contracting Parties.

(d) Contracting Parties shall provide to other Contracting Parties and to SCAR at least 30 days in advance of departure from their home ports, information on proposed sealing expeditions.

7. *Sealing methods*

(a) SCAR is invited to report on methods of sealing and to make recommendations with a view to ensuring that the killing or capturing of seals is quick, painless and efficient. Contracting Parties, as appropriate, shall adopt rules for their nationals and vessels under their respective flags engaged in the killing and capturing of seals, giving due consideration to the views of SCAR.

(b) In the light of the available scientific and technical data, Contracting Parties agree to take appropriate steps to ensure that their nationals and vessels under their respective flags refrain from killing or capturing seals in the water, except in limited quantities to provide for scientific research in conformity with the objectives and principles of this Convention. Such research shall include studies as to the effectiveness of methods of sealing from the viewpoint of the management and humane and rational utilization of the Antarctic seal resources for conservation purposes. The undertaking and the results of any such scientific research programme shall be communicated to SCAR and the Depositary which shall transmit them to the Contracting Parties.

Signatures and Ratifications

State	Date of signature	Date of deposit of ratification
Argentine Republic	9 June 1972	7 Mar. 1978
Australia	5 Oct. 1972	
Belgium	9 June 1972	9 Feb. 1978
Chile	28 Dec. 1972	7 Feb. 1980
France	19 Dec. 1972	19 Feb. 1975
Japan	28 Dec. 1972	28 Aug. 1980
New Zealand	9 June 1972	
Norway	9 June 1972	10 Dec. 1973
South Africa	9 June 1972	15 Aug. 1972
Union of Soviet Socialist Republics	9 June 1972	8 Feb. 1978
United Kingdom	9 June 1972	10 Sept. 1974
United States of America	28 June 1972	28 Dec. 1976
Convention entered into force	11 March 1978	

UN Resolution 38/77

15 December 1983

The General Assembly,

Having considered the item entitled 'Question of Antarctica',

Conscious of the increasing international awareness of and interest in Antarctica,

Bearing in mind the Antarctic Treaty[1] and the significance of the system it has developed,

Taking into account the debate on this item at the thirty-eighth session of the General Assembly,

Convinced of the advantages of a better knowledge of Antarctica,

Affirming the conviction that, in the interest of all mankind, Antarctica should continue to be used exclusively for peaceful purposes, and that it should not become the scene or object of international discord,

Recalling the relevant paragraphs of the Declaration of Heads of State or Government of the Non-Aligned Countries at their Seventh Conference, held at New Delhi from 7 to 12 March 1983,[2]

1. *Requests* the Secretary-General to prepare a comprehensive, factual study on all aspects of Antarctica, taking fully into account the Antarctic Treaty system and other relevant factors;

2. *Also requests* the Secretary-General to seek the views of all Member States in the preparation of the study;

3. *Requests* those States conducting scientific research in Antarctica, other interested States, the relevant specialized agencies, organs, organizations and bodies of the United Nations system, and relevant international organizations having scientific or technical information on Antarctica, to lend the Secretary-General whatever assistance he may request for the purpose of carrying out the study;

4. *Requests* the Secretary-General to report to the General Assembly at its thirty-ninth session;

5. *Decides* to include in the provisional agenda of its thirty-ninth session the item entitled 'Question of Antarctica'.

References

1. For the text of the Treaty, see *The United Nations and Disarmament*: 1945–1970 (United Nations publication, Sales No. 70. IX.l), appendix IV.

2. See A/38/132-S/15675 and Corr. 1 and 2.

UN Resolution 39/152

17 December 1984

The General Assembly,

Recalling its resolution 38/77 of 15 December 1983,

Having considered the item entitled 'Question of Antarctica',

Noting the study on the 'Question of Antarctica',[1]

Conscious of the increasing international awareness of and interest in Antarctica,

Bearing in mind the Antarctic Treaty[2] and the significance of the system it has developed,

Taking into account the debate on this item at its thirty-ninth session,[3]

Convinced of the advantages of a better knowledge of Antarctica,

Affirming the conviction that, in the interest of all mankind, Antarctica should continue forever to be used exclusively for peaceful purposes and that it should not become the scene or object of international discord,

Recalling the relevant paragraphs of the Declaration of Heads of State or Government of the Non-Aligned Countries at their Seventh Conference, held at New Delhi from 7 to 12 March 1983,[4]

1. *Expresses its appreciation* to the Secretary-General for the study entitled 'Question of Antarctica',[5]

2. *Decides* to include in the provisional agenda of its fortieth session the item entitled 'Question of Antarctica'.

References

1. A/39/583 (Part I) and Corr. 1 and (Part II, vols. I–III) and Corr. 1.

2. United Nations, *Treaty Series*, vol. 402, No. 5778, p. 72.

3. *Official Records of the General Assembly, Thirty-Ninth Session, First Committee*, 50th and 52nd to 55th meetings.

4. A/38/132-S/15675 and Corr. 1 and 2, annex, sect. III, paras. 122 and 123.

5. A/39/583 (Part I) and Corr. 1 and (Part II, vols. I-III) and Corr. 1.

UN Resolution 40/156

16 December 1985

Part A

The General Assembly,

Recalling its resolutions 38/77 of 15 December 1983 and 39/152 of 17 December 1984,

Having considered the item entitled 'Question of Antarctica',

Welcoming the increasing international awareness of the interest in Antarctica,

Bearing in mind the Antarctic Treaty[1] and the significance of the system it has developed,

Taking into account the debate on this item at its fortieth session,[2]

Convinced of the advantages of a better knowledge of Antarctica,

Affirming the conviction that, in the interest of all mankind, Antarctica should continue forever to be used exclusively for peaceful purposes and that it should not become the scene or object of international discord,

Recalling the relevant paragraphs of the Economic Declaration adopted at the Seventh Conference of Heads of State or Government of Non-Aligned Countries, held at New Delhi from 7 to 12 March 1983,[3] and of the Final Political Declaration of the Conference of Foreign Ministers of Non-Aligned Countries held at Luanda from 4 to 7 September 1985,[4] as well as the resolution on Antarctica adopted by the Council of Ministers of the Organization of African Unity at its forty-second ordinary session, held at Addis Ababa from 10 to 17 July 1985,[5]

Conscious of the significance of Antarctica to the international community in terms, *inter alia*, of international peace and security, economy, environment, scientific research and meteorology,

Recognizing, therefore, the interest of mankind as a whole in Antarctica,

Bearing in mind the coming into force of the United Nations Convention on the Law of the Sea,[6]

Noting once again with appreciation the study on the question of Antarctica,[7]

Convinced that it would be desirable to examine further certain issues affecting Antarctica,

1. *Requests* the Secretary-General to update and expand the study on the question of Antarctica by addressing questions concerning the availability to the United Nations of information from the Antarctic Treaty Consultative Parties on their respective activities in and their deliberations regarding Antarctica, the involvement of the relevant specialized agencies and intergovernmental organizations in the Antarctic Treaty system and the significance of the United Nations Convention on the Law of the Sea in the southern ocean;

2. *Requests* the Secretary-General to seek the co-operation of all Member States and the relevant specialized agencies, organs, organizations and bodies of the United Nations system, as well as the relevant intergovernmental and non-governmental bodies, in the preparation of the study by inviting them to transmit, as appropriate, their views and any information they may wish to provide;

3. *Requests* the Secretary-General to submit the study to the General Assembly at its forty-first session;
4. *Decides* to include in the provisional agenda of its forty-first session the item entitled 'Question of Antarctica'.

Part B

The General Assembly,

Recalling its resolutions 38/77 of 15 December 1983 and 39/152 of 17 December 1984,

Having considered the item entitled 'Question of Antarctica',

Recalling the relevant paragraphs of the Economic Declaration adopted at the Seventh Conference of Heads of State or Government of Non-Aligned Countries, held at New Delhi from 7 to 12 March 1983,[3] and of the Final Political Declaration of the Conference of Foreign Ministers of Non-Aligned Countries held at Luanda from 4 to 7 September 1985,[4] as well as the resolution on Antarctica adopted by the Council of Ministers of the Organization of African Unity at its forty-second ordinary session, held at Addis Ababa from 10 to 17 July 1985,[5]

Recognizing that the management, exploration and use of Antarctica should be conducted in accordance with the purpose and principles of the Charter of the United Nations and in the interest of maintaining international peace and security and of promoting international co-operation for the benefit of mankind as a whole,

Aware that negotiations are in progress among the Antarctic Treaty Consultative Parties, with the non-Consultative Parties as observers, to which other States are not privy, with a view to establishing a régime regarding Antarctic minerals,

1. *Affirms* that any exploitation of the resources of Antarctica should ensure the maintenance of international peace and security in Antarctica, the protection of its environment, the non-appropriation and conservation of its resources and the international management and equitable sharing of the benefits of such exploitation;
2. *Invites* the Antarctic Treaty Consultative Parties to inform the Secretary-General of their negotiations to establish a régime regarding Antarctic minerals;
3. *Requests* the Secretary-General to submit to the General Assembly for consideration at its forty-first session a report containing the replies received from Consultative Parties;
4. *Decides* to include in the provisional agenda of its forty-first session the item entitled 'Question of Antarctica'.

Part C

The General Assembly,

Having considered the item entitled 'Question of Antarctica',

Noting with regret that the racist *apartheid* régime of South Africa, which has been suspended from participation in the General Assembly of the United Nations, is a Consultative Party to the Antarctic Treaty,[1]

Recalling the interest of African States in Antarctica as shown by the resolution adopted by the Council of Ministers of the Organization of African Unity at its forty-second ordinary session, held at Addis Ababa from 10 to 17 July 1985,[5]

Recalling further that the Antarctic Treaty is, by its terms, intended to further the purposes and principles embodied in the Charter of the United Nations,

1. *Views with concern* the continued status of the *apartheid* régime of South Africa as a Consultative Party to the Antarctic Treaty;
2. *Urges* the Antarctic Treaty Consultative Parties to exclude the racist *apartheid* régime of South Africa from participation in the meetings of the Consultative Parties at the earliest possible date;
3. *Invites* the States parties to the Antarctic Treaty to inform the Secretary-General on the actions taken regarding the provisions of the present resolution.

References

1. United Nations, *Treaty Series*, vol. 402, No. 5778, p. 72.
2. *Official Records of the General Assembly, Fortieth Session, First Committee*, 48th to 55th meetings.
3. A/38/132-S/15675 and Corr. 1 and 2, annex, sect. III, paras. 122 and 123.
4. A/40/854-S/17610, annex I, sect. V.
5. A/40/666, annex II, resolution CM/Res. 988 (XLII).
6. *Official Records of the Third United Nations Conference on the Law of the Sea*, vol. XVII (United Nations publication, Sales No. E.84.V.3), document A/CONF.62/122.
7. A/39/583 (Part I) and Corr. 1, 2 and 3 and A/39/583 (Part II) and Corr. 1, vols. I–III.

Antarctic Treaty Member States and their affiliations

Country	Original signatory	ATCS[1]	UNSC[2]	G77	NAM	Year[3]
United Kingdom	X	X	X			1960
South Africa	X	X				1960
Belgium	X	X				1960
Japan	X	X				1960
USA	X	X	X			1960
Norway	X	X				1960
France	X	X	X			1960
New Zealand	X	X				1960
USSR	X	X	X			1960
Poland		X				1961
Argentina	X	X		X	X	1961
Australia	X	X				1961
Chile	X	X		X		1961
Czechoslovakia						1962
Denmark						1965
Netherlands						1967
Romania				X		1971
German Democratic Republic						1974
Brazil		X		X		1975
Bulgaria						1978
Federal Republic of Germany		X				1979
Uruguay		X		X		1980
Papua New Guinea				X		1981
Italy						1981
Peru				X	X	1981
Spain						1982
China		X	X			1983
India		X		X	X	1983
Hungary						1984
Sweden						1984
Finland						1984
Cuba				X	X	1984

1. Antarctic Treaty Consultative Party status.
2. Permanent Member of the UN Security Council.
3. The date given is the date of ratification for the 12 Original Signatories, and the date of accession or succession for other States.

Members of the UN group of 77 and the Non-Aligned Movement

States given in capitals are co-sponsors of UN Resolutions on Antarctica, States given in italics have acceded to the Antarctic Treaty. Argentina, Brazil, Chile, India and Uruguay are Consultative Parties.

Afghanistan	Ecuador	Malawi
Algeria	Egypt	MALAYSIA
Angola	El Salvador[1]	Maldives
ANTIGUA & BARBUDA[1]	Equatorial Guinea	MALI
Argentina	Ethiopia	Malta
Bahamas	Fiji[1]	Mauritania
Bahrain	Gabon	MAURITIUS
BANGLADESH	Gambia	Mexico[1]
Barbados	Ghana	Morocco
Belize	Grenada	Mozambique
Benin	Guatemala[1]	Nepal
Bhutan	Guinea	Nicaragua
Bolivia	Guinea-Bissau	Niger
Botswana	Guyana	NIGERIA
Brazil[1]	Haiti[1]	OMAN
BRUNEI[1]	Honduras[1]	PAKISTAN
Burkina	*India*	Palestine Liberation
Burma[1]	INDONESIA	Front[1]
Burundi	Iran	Panama
Cambodia	Iraq	*Papua New Guinea*[1]
CAMEROON	Ivory Coast	Paraguay[1]
Cape Verde	Jamaica	*Peru*
Central African Rep.	Jordan	PHILIPPINES[1]
Chad	Kenya	Qatar
Chile[1]	Korea, Rep. of[1,2]	*Romania*[1]
Colombia	Korea, Democratic	RWANDA
Comoros	People's Rep.[2]	St Kitts & Nevis[1]
Congo	Kuwait	St Lucia
Costa Rica[1]	Laos	St Vincent[1]
Cuba	Lebanon	Samoa (Western)[1]
Cyprus	Lesotho	Sao Tome & Principe
Djibouti	Liberia	Saudi Arabia
Dominica[1]	Libya	Senegal
Dominican Republic[1]	Madagascar	Seychelles

continued

Sierra Leone	Tanzania	Vanuatu
Singapore	Thailand[1]	Venezuela[1]
Solomon Islands[1]	Togo	Vietnam
Somalia	Tonga[1,3]	North Yemen
SRI LANKA	Trinidad & Tobago	South Yemen
Sudan	Tunisia	Yugoslavia
Surinam	Uganda	Zaire
Swaziland	United Arab Emirates	Zambia
Syria	*Uruguay*[1]	Zimbabwe

1. Not a member of the NAM.
2. Observer status at the UN.
3. Not a member of the UN.

Notes and references

Chapter 1

1. The question of Antarctica was referred to the First Committee of the General Assembly: debates in that Committee led to the adoption, in successive years, of General Assembly Resolutions 38/77, 39/152 and 40/156. See also the Study of the Secretary-General on the Question of Antarctica, UN Doc. A/39/583 (1984). An earlier attempt by India to put Antarctica on the agenda of the General Assembly, in 1958, was not pressed: *UN Yearbook*, 1958, pp. 109–10.
2. 402 United Nations *Treaty Series* 71; United Kingdom *Treaty Series* No. 97 (1961).
3. UN Secretary-General's Study on the Question of Antarctica, UN Doc. A/39/583 (1984), paragraph 42 (although that paragraph incorrectly puts the place in question – Hope Bay – in the South Orkney Islands: it is in fact on the Antarctic Peninsula). See paragraphs 35–43 generally for a summary of various other incidents occasioning tension or conflict in the Antarctic between 1945 and the conclusion of the Antarctic Treaty in 1959.
4. See generally on claims to territorial sovereignty in Antarctica, Auburn, *Antarctic Law and Politics*, C. Hurst & Co., London (1982), pp. 5–47; UN Secretary-General's Study on the Question of Antarctica, UN Doc. A/39/583 (1984), paragraphs 11–57. The basis for the United Kingdom's sovereignty over part of Antarctica is set out in its Applications of May 1955 to the ICJ Instituting Proceedings against Argentina and Chile (*ICJ Pleadings, Antarctica Cases*).
5. The IGY consisted essentially of national scientific programmes, co-ordinated through a Special Committee for the IGY established by the International Council of Scientific Unions. See generally Auburn, *op. cit.* in note 4, pp. 86–93, and also Shapley, *The Seventh Continent: Antarctica in a Resource Age*, Resources for the Future Inc., Washington DC (1985) pp. 82–8.
6. See Auburn, *op. cit.* in note 4, pp. 89–91.
7. Originally the Special Committee on Antarctic Research. On SCAR generally, see Zumberge in *Antarctic Treaty System: An Assessment*, National Research Council Proceedings of the Polar Research Board Commission on the Physical Sciences, Mathematics and Resources National Academy Press, Washington DC (1986) pp. 153–68.
8. See generally, *Handbook of the Antarctic Treaty System* (4th edn 1985) (hereafter cited as '*Handbook*').
9. *Ibid.* p. 2102; Antarctic Treaty Act 1967, Schedule 2.
10. United Kingdom *Treaty Series* No. 11 (1973); see Appendix II for text.
11. United Kingdom *Treaty Series* No. 48 (1982); see Appendix II for text.

12. In a broader sense the Antarctic Treaty System may be regarded as comprising also certain elements which do not flow directly from the Antarctic Treaty, such as the International Convention for the Regulation of Whaling 1946 (UK *Treaty Series*, No. 5 (1949)), and SCAR, on which see above, note 7.
13. Argentina, Australia, Chile, France, New Zealand, Norway and the United Kingdom.
14. Rec. XI-1 adopted at the 11th Antarctic Treaty Consultative Meeting, Buenos Aires, 1981 (*Handbook*, p. 1512).
15. Report of the 12th Antarctic Treaty Consultative Meeting, Canberra, 1983, paragraphs 38–41 (*Handbook*, pp. 6109–10); Rec. XIII-15 adopted at the 13th Antarctic Treaty Consultative Meeting, Brussels, 1985. See also Rec. XIII-2 adopted at that Meeting and paragraphs 20–21 and 73–75 of the Report of that Meeting as regards the possibility of inviting observers from appropriate international organisations.
16. See Rule 23 of the Rules of Procedure of Consultative Meetings (*Handbook*, p. H6) as regards the approval of recommendations. The approval of the final report, and all matters of procedure, require only a majority decision: Rules 24 and 20. Measures recommended to Governments by Consultative Meetings become effective only when approved by all Consultative Parties entitled to attend the Meeting in question: Antarctic Treaty, Article IX.4.
17. Article XII.2(a).
18. See also the seventh principle in the Declaration on Principles of International Law concerning Friendly Relations and Co-operation among States in accordance with the Charter of the United Nations, GA Res. 2625(XXV).
19. Charter of the United Nations, Article 103.

Chapter 2

1. Rec. III–VII; Article X.
2. Convention on the Conservation of Antarctic Marine Living Resources, Article V.
3. UN General Assembly Resolution 3201 (S–VI), 1 May 1974.
4. UN General Assembly Resolution 3281 (XIX), 12 December 1974.
5. UN General Assembly Resolution 2574 (XXIV), 15 December 1969.
6. For example, Declaration by the Ministers of Foreign Affairs of the States Members of the Group of 77 at New York, 23 October 1977, UN document A/34/611.
7. *cf.* Article 86 of the UN Charter on composition of the Trusteeship Council and the UN Convention on the Law of the Sea, Article 161.1 on composition of the Council.
8. *cf.* Statute of the IAEA, Article VI(A).2(a) and Article 161.1(e) of the UN Convention on the Law of the Sea.

Chapter 3

1. This was certainly the case in 1984.
2. Taken from publicity leaflet entitled *Greenpeace Statement on the United Nations* issued by Greenpeace International on 12.9.84 at the time of the CCAMLR III meeting in Hobart.
3. Extract from the 'Alternative Principles' as set in Paris on 23.9.85 in newsletter *Eco*, Vol. XXXIII, No. 1.
4. Mike Donoghue in *Campaigners Antarctic Notes, No. 35* Greenpeace (Auckland), 3 October 1985.

5. These procedures were derived from Benninghoff, W. S. and Bonner, W. N., *Man's Impact on the Antarctic Environment*, SCAR, Cambridge (1985).
6. Mitchell B., 'Antarctic Politics and Marine Resources: Critical Choices for the 1990s'. In Alexander, L. M. and Hanson, L. C. (eds.), *Proceedings from the 8th Annual Conference of the Centre for Ocean Management Studies*, 17–20 June 1984, University of Rhode Island, 1985.

Chapter 4

1. Craddock, C. (ed.), *Antarctic Geoscience, Symposium on Antarctic Geology and Geophysics, Madison, Wisconsin, 22–27 August 1977*, University of Wisconsin Press, Madison, 1172 pp. (1982).
2. Drewry, D. J. (ed.), *Antarctica: glaciological and geophysical folio*, Scott Polar Research Institute, Cambridge (1983).
3. Lorius, C., *et al.*, 'A 150,000 year climate record from Antarctic ice', *Nature*, vol. 416, pp. 591–6 (1985).
4. Peel, D. A., 'Antarctic ice: the frozen time capsule'. *New Scientist*, vol. 98, pp. 476–9 (1983).
5. Rubin, M. J., 'The Antarctic and the weather'. *Scientific American*, vol. 207, pp. 84–94 (1962).
6. Farman, J. C., Gardiner, B. G. and Shanklin, J. D., 'Large losses of total ozone in Antarctica reveal seasonal ClO_x/NO_x interactions'. *Nature*, vol. 315, pp. 207–10 (1985).
7. Dudeney, J. R., 'The ionosphere, a view from the Pole', *New Scientist*, vol. 91, pp. 714–17 (1981).
8. Rycroft, M. J., 'A view of the upper atmosphere from Antarctica', *New Scientist*, vol. 108, pp. 44–51 (1985).
9. Laws, R. M., 'Ecology of the Southern Ocean', *American Scientist*, vol. 73, pp. 26–40 (1985).
10. Siegfried, W. R., Condy, P. R. and Laws, R. M. (eds.), *Antarctic Nutrient Cycles and Food Webs, Fourth SCAR Symposium on Antarctic Biology, Wilderness, South Africa, 12–16 September 1983*, Springer-Verlag, New York, 700 pp. (1985).

Chapter 5

1. Hardin, G., 'The Tragedy of the Commons', *Science*, vol. 162, pp. 1243–7 (1968).
2. Mitchell, B., and Sandbrook, R., *Management of the Southern Ocean*, International Institute for Environment and Development, London and Washington (1979).
3. Birnie, P., *Marine Policy*, vol. 10, No. 1, pp. 62–6 (1986).
4. Benninghoff, W. S. and Bonner, W. N., *Man's Impact on the Antarctic Environment*, SCAR, Cambridge (1985).
5. SCAR Group of Specialists, *Antarctic Environmental Implications of Possible Mineral Exploration and Exploitation, Report No. 1*, SCAR, Cambridge (1981).
6. SCAR Group of Specialists, *Antarctic Environmental Implications of Possible Mineral Exploration and Exploitation, Report No. 2*, SCAR, Cambridge (1983).
7. Mitchell, B. and Tinker, J., *Antarctica and its resources*, International Institute for Environment and Development, London and Washington, (1970).

Chapter 6

1. Wilsher, P. and Janmohamed, P., 'The New Continents, Part III: The Polar Regions', *Sunday Times Magazine*, pp. 44–7, 8 December 1985.

2. Thomson, M. and Swithinbank, C. W., 'The Prospects for Antarctic Minerals', *New Scientist*, vol. 107, No. 1467, pp. 31–5 (1985).
3. Sugden, D., *Arctic and Antarctic: a modern geographical synthesis*, Blackwell, Oxford (1982).
4. Elliot, D. H., 'Physical Geography – Geological Evolution'. In Bonner, W. N. & Walton, D. W. H. (eds.), *Key Environments: Antarctica*, Pergamon Press, Oxford (1985).
5. Behrendt, J. C. (ed.), 'Petroleum and Mineral Resources of Antarctica', *USGS Geological Circular*, No. 909 (1983).
6. Behrendt, J. C., 'Scientific Studies on the Question of Petroleum Resources of Antarctica'. In Splettstoesser, J. F. (ed.), *Mineral Resource Potential of Antarctica*, University of Texas Press (in press).
7. Ford, A. B., 'The Dufek Intrusion in Antarctica and a survey of its minor metals and possible resources'. In Behrendt, J. C. (ed.), 'Petroleum and Mineral Resources of Antarctica', *USGS Geological Circular*, No. 909 (1983).
8. de Wit, M., *Minerals and Mining in Antarctica*, Oxford University Press (1985).
9. Neshyba, S., 'On the size and distribution of Antarctic icebergs', *Cold Regions Science and Technology*, vol. 1, pp. 241–8 (1980).
10. Sanderson, R. J. O. and Larminie, F. G., 'Offshore oil development in polar regions', unpublished contribution, 15th Pacific Science Congress, Dunedin, New Zealand, (1983).

Chapter 7

1. Executive Summary of the *Report of the United States Observers Team in Antarctica, 1983.*
2. *Foreign Report*, 20 January 1983.
3. *The Mershon Center Quarterly Report*, vol. 8, No. 4, p. 6, Spring 1984.
4. *Aussenpolitik*, No. 1, p. 84, Spring 1984.
5. FCO Background Brief 'Antarctic Treaty, 25th Anniversary', Foreign and Commonwealth Office, London, November 1984.
6. Luard, E., 'Who owns the Antarctic?', *Foreign Affairs*, vol. 62, No. 5, Summer 1984.
7. *Bulletin of the Atomic Scientists*, vol. 40, No. 6, pp. 31–2, June/July 1984.
8. Gottfried, K. and Lebow, R. N., 'Weapons in Space', *Daedalus*, vol. 1, p. 148, Spring 1984.
9. Unpublished paper presented to the 26th Annual Convention of the International Studies Association, Washington DC, 7 March 1983, pp. 6–7.
10. *Journal of International Law*, p. 232, Spring 1985.
11. *Op. cit.*, p. 282.

Index